From Microstructure Investigations to Multiscale Modeling

Series Editor
Félix Darve

From Microstructure Investigations to Multiscale Modeling

Bridging the Gap

Edited by

Delphine Brancherie
Pierre Feissel
Salima Bouvier
Adnan Ibrahimbegović

iSTE

WILEY

First published 2017 in Great Britain and the United States by ISTE Ltd and John Wiley & Sons, Inc.

ISTE Ltd
27-37 St George's Road
London SW19 4EU
UK

www.iste.co.uk

John Wiley & Sons, Inc.
111 River Street
Hoboken, NJ 07030
USA

www.wiley.com

Library of Congress Control Number: 2017953125

British Library Cataloguing-in-Publication Data
A CIP record for this book is available from the British Library
ISBN 978-1-78630-259-5

MIX
Paper from responsible sources
FSC® C013604

Contents

Preface

Mechanical behaviors of materials are highly influenced by their microstructures. Therefore, progress in material science aims at understanding and modeling the link between the microstructure and the material behavior at different scales. This book comprises contributions from eminent researchers in the field of computational and experimental material modeling. The book focuses on experimental techniques, modeling approaches and computational strategies to understand and predict the behavior of materials in relation to its architecture and microstructure at different scales. Special attention is paid to the coupling of experimental techniques with advanced modeling tools, numerically or analytically.

The first four chapters are dedicated to the reconstruction of representative volume element (RVE) for different kinds of materials to study the mechanical behavior at the macro-scale. Advanced experimental techniques along with dedicated numerical and analytical tools are presented to efficiently analyze and represent the microstructural features. These tools are used to study synthetic materials, the key properties of RVEs, and to construct the behavior at the macro-scale through homogenization. The role of the randomness of the microstructure in the macro-scale behavior is also investigated, and stochastic dedicated tools are presented.

The following three chapters focus on complex mechanical behavior modeling at the macro-scale. Different modelization and simulation techniques are presented. These chapters discuss how the applications considered above enable the use and adaptation of numerical tools to analyze complex behaviors. The macro phenomenological models are based on a better understanding and modeling of key phenomena at the micro-scale, including multi-physics. The influence of the parameters of micro-scale models is analyzed in detail with respect to the macro-scale behavior.

The increasing complexity of models brings challenging issues related to identification and simulation. Due to these issues, models must be identified from complex experiments that are monitored richly, to imply the large amount of data. Dedicated identification strategies must be developed based on simulation, thereby requiring model reduction techniques. These reduction techniques will also be a key tool for large-scale (in terms of CPU time) predictive simulation, and new trends in data-driven simulations take advantage of the experimental data to propose a new modeling paradigm. The last two chapters focus on these issues.

This book is a collection of selected papers from the invited lectures presented at the 9th US–France symposium: "From microstructure observations to multi-scale modeling of deformation mechanisms and interfaces". This symposium was held in Compiègne in June 2016 under the auspices of the International Center for Applied Computational Mechanics, Compiègne, France, 1–3 June 2016.

We would like to thank all the ICACM participants for their lively exchanges, especially the authors of the chapters for their contributions.

Delphine Brancherie
Pierre Feissel
Salima Bouvier
Adnan Ibrahimbegović
September 2017

1

Synchrotron Imaging and Diffraction for *In Situ* 3D Characterization of Polycrystalline Materials

1.1. Introduction

The last few years have seen material science progressing rapidly into three-dimensional (3D) characterization at different scales (e.g. atom probe tomography [PHI 09], transmission electron microscopy tomography [WEY 04], automated serial sectioning tomography [UCH 12, ECH 12] and X-ray tomography [MAI 14]). A wealth of 3D data sets can now be obtained with different modalities, allowing the 3D characterization of phases, crystallography, chemistry, defects or damage and in some cases strain fields.

In the last 10 years, one particular focus of the 3D imaging community (like 2D in its time with the advent of EBSD characterization) has been on obtaining reliable three-dimensional grain maps. As most structural materials are polycrystalline and the mechanical properties are determined by their internal microstructure, this is a critical issue. There has been considerable effort to develop characterization techniques at the mesoscale, which can image

Chapter written by Henry PROUDHON.

typically 1 mm^3 of material with a spatial resolution in the order of micrometers.

Among 3D characterization, an important distinction exists between destructive and non-destructive techniques. Serial sectioning relies on repeated 2D imaging (which may include several modalities) of individual slices, where a thin layer of material is removed between each observation (see Figure 1.1(a,b)). The material removal can be achieved via mechanical polishing [ROW 10], ion [DUN 99, JIR 12] or femtosecond laser ablation [ECH 15] in a dedicated scanning electron microscope (SEM). Considerable progress has been made in this line in the last decade, bringing not only high-quality measurements in 3D of grain sizes and orientations but also detailed grain shapes and grain boundary characters. The most serious threat of serial sectioning is, however, the destruction of the sample.

In parallel, the advent of third-generation synchrotrons worldwide, with ESRF at the forefront, brought hard X-rays, with their high penetrating power, to the structural material science community. X-ray computed tomography (CT) rapidly developed as a key observation tool, allowing the non-destructive bulk evaluations of all types of materials [MAI 14]. This made the *in situ* study of damage possible using specifically designed stress rigs [BUF 10]. Unfortunately, CT imaging relies on absorption and phase contrasts and remains blind to crystal orientation. Accessing crystallographic information in the bulk of polycrystalline specimens (average orientation per grain) was subsequently achieved using the high penetrating power of hard X-rays and leveraging diffraction contrast. The pioneering work of Poulsen took advantage of high-brilliance synchrotron sources to study millimeter-sized specimens by tracking the diffraction of each individual crystal within the material volume while rotating the specimen over 360°. This led to the development of 3DXRD [POU 04] and its several grain mapping variants (DCT [LUD 08], HEDM [LIE 11], DAGT [TOD 13]). Among them, the near-field variant called diffraction contrast tomography (DCT, see Figure 1.1(c)) will be detailed in section 1.2.7.

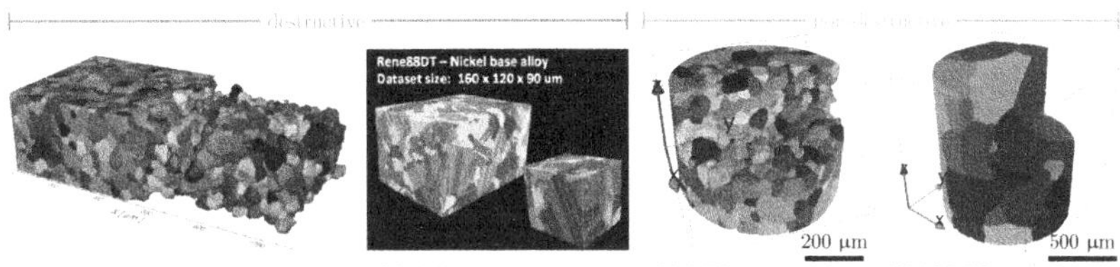

(a) β-Ti reconstruction by EBSD serial sectioning (b) Tribeam reconstruction of a Rene88 sample (c) DCT reconstruction of a pure Ti sample (d) labDCT reconstruction of a test AlLi sample

Figure 1.1. *Various examples of grain mapping techniques (spatial resolution indicated in brackets): (a) serial sectioning through mechanical polishing (0.45 μm in the XY plane, 1.48 μm in the Z direction) [ROW 10], (b) tribeam (laser ablation, 0.75 μm) [ECH 15], (c) DCT near-field imaging (1.4 μm) [PRO 16a], (d) labDCT reconstruction with grain shape and orientations (5 μm, courtesy of Xnovotech). For a color version of the figure, see www.iste.co.uk/brancherie/microstructure.zip*

This chapter is organized as follows: the fundamentals of 3D X-ray characterization of structural materials are reviewed in section 1.2 from pure absorption contrast to diffraction contrast tomography. Then, a recently developed stress rig adapted for DCT is presented in section 1.3. Finally, section 1.4 details how to use experimental 3D grain maps in finite-element crystal plasticity calculations.

1.2. 3D X-ray characterization of structural materials

1.2.1. *Early days of X-ray computed tomography*

As most materials are opaque to optic light, people have long relied on the observation of object surfaces and/or destructive cut to assess various evolutions of physical phenomenon or diagnostics. With the advent of X-rays in the 20th Century, both easy to produce and capable of penetrating significantly through matter, a large number of works have been devoted to develop a technique that would allow to *see the invisible*.

The theoretical bases of tomography were laid down by J. Radon in 1917 (see section 1.2.4), and the first part of the 20th Century saw the development of what we call today conventional tomography. This allowed to record sectional image through a body by moving X-ray source and the film in opposite directions during the exposure. However, the images remained difficult to interpret.

The development of computer power in the second part of the 20th Century let G. Hounsfield build the first computed tomography (CT) scanner in 1971. At that time, the spatial resolution achieved was about 3 mm and the resolution of the images produced after reconstruction by a Data General Nova 16 bits minicomputer was 80 × 80 pixels. The scanner was installed in an English hospital, where the first brain scan was performed on a patient.

In the United States, A. Cormack independently developed a similar system, and both of them received the Nobel Prize in Medicine in 1979 for their invention. Since then, CT scanning has been used in medical practice as a powerful tool to help diagnostic.

1.2.2. *X-ray absorption and Beer Lambert's law*

In the 1–100 keV[1] energy range typically used for material science investigations, the interactions of an X-ray photon with matter are largely dominated by photoelectric absorption and elastic Thomson scattering. Unless specific conditions for diffraction are fulfilled (see section 1.2.7), the main contrast in a radiograph is due to absorption.

This phenomenon of primary importance non-restricted to X-rays was first observed by A. Becquerel in 1839 and later explained by A. Einstein in his famous 1905 paper using quanta of light (that we now call photon) conceptualized by M. Planck in 1900. This essentially allowed to explain that the electromagnetic field is quantified and can interact with electrons of the outer shell of an atom with the net result of being absorbed and that a photoelectron is emitted isotropically (fluorescence, Figure 1.2(a)). The absorption is thus related to the probability of the X-ray photon penetrating some material for being absorbed and will therefore strongly depend on the

1 Above 100 keV, inelastic Compton scattering becomes significant.

incident photon energy E and the number of electrons or equivalently the atomic number Z of the material.

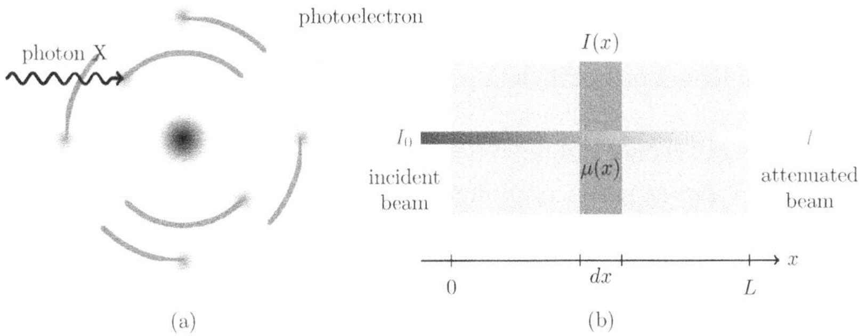

Figure 1.2. *(a) Absorption of an X-ray photon by an atom, and emission of a photoelectron (fluorescence); (b) attenuation of an X-ray beam of intensity I_0 by a piece of material characterized by its absorption coefficient $\mu(x)$*

To quantify this effect, the linear absorption coefficient (μ) is introduced. By definition, μdx is the attenuation of the beam through a very small section of material of thickness dx lying at the distance x from the surface (Figure 1.2(b)). Thus, we have:

$$-dI = I(x)\mu dx \quad \text{or} \quad \frac{dI}{I(x)} = -\mu dx \qquad [1.1]$$

Using the conditions $I(x = 0) = I_0$ and $I(x = L) = I$ leads to the Beer–Lambert law:

$$\frac{I}{I_0} = \exp -(\mu L) \qquad [1.2]$$

If the attenuation of the object is not uniform, the Beer–Lambert law can be derived similarly with an integral:

$$\ln \frac{I_0}{I} = \int_0^L \mu(x)dx \qquad [1.3]$$

It is experimentally observed that the μ scale is approximately like Z^4/E^3 and displays strong jumps in energy (called absorption edges) when crossing the corresponding atom–electron binding energies (for instance, K edge is 1.56 keV for Al and 88.0 keV for Pb). In practice, the mass attenuation coefficient $\mu_\rho = \mu/\rho$ has been precisely measured and tabulated by crystallographers for all types of materials [SEL 93]. Reliable data sets are publicly available, such as the ones published by the NIST[2]. An example is plotted for both Al and Pb in Figure 1.3, which clearly shows why lead is used as shielding for X-rays: at 20 keV and for a thickness of 1 mm, the radiation is totally absorbed by the Pb sheet, whereas 42% of the photons have crossed in the case of Al.

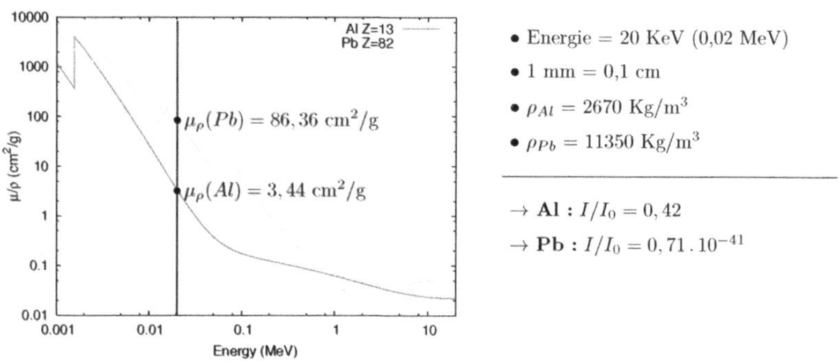

Figure 1.3. *X-ray attenuation through 1 mm of metal: Al vs Pb, note the absorption edges visible on the evolution of μ_ρ. For a color version of the figure, see www.iste.co.uk/brancherie/microstructure.zip*

1.2.3. *X-ray detection*

Once the radiation has been attenuated by the sample, it must be recorded in some way to form an image. Silver-coated photosensitive films have long been used for X-ray experiments (and are still in used in many cases for radiology). All detectors now produce digital data of the recorded signal. Three main X-ray detection technologies producing 2D images are usually distinguished:

2 http://www.nist.gov/pml/data/xraycoef/

– *Image plate detectors* now advantageously replace photosensitive films (Figure 1.4(a)). The detector is covered by a phosphor layer, which is quantitatively sensitive to short-wavelength radiation. When exposed to X-rays, a two-dimensional image is created on the phosphor screen. To retrieve the recorded image, the screen has to be exposed again to visible light (typically using a laser) and the signal is then processed to produce the final digital image.

– *Pixel detectors* have been developed recently thanks to the progress of the microelectronics industry. Silicium chips, in which each pixel has its own amplifier, are directly exposed to X-rays (Figure 1.4(b)). Mostly used for moderate X-ray energies, they feature zero noise, very high sensitivity (single photon counting) and very short readout time (in the order of 1 ms).

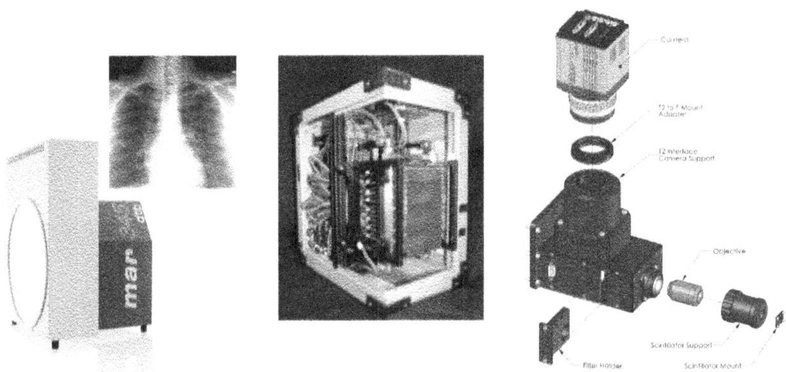

Image plate from marXperts (a) | Xpad pixel detector from imXpad (b) | CCD + optical system from Optique Peter (c)

Figure 1.4. *Various X-ray detection systems exploiting different technologies*

– *Scintillator+CCD* systems are now quite common for X-ray imaging experiments. The scintillator (usually made of Gadox for gadolinium oxysulfide) screen exposed to X-rays convert the X-ray photons into visible light photons, which thanks to an optical system are collected by a square CCD chip to form the image. One advantage is that the scintillator and the optics can be tuned to achieve a range of spatial resolution (Figure 1.4(c)).

1.2.4. *Radon's transform and reconstruction*

Radon showed mathematically that a function could be reconstructed from an infinite set of its projections [RAD 17], which is known as the Radon theorem. Here, we shall consider the simple case of parallel geometry corresponding to most synchrotron chromatography setups (see section 1.2.5). The principle extends to more complex geometry (cone and fan beam, for instance, see [GRA 02]). Given that the object is represented in the plane by a function of its physical properties $f(x, y)$ and a projection angle θ, the Radon transform $\mathcal{R}f_\theta(t)$ is the integral of f along the line D_t of direction $(-\sin\theta, \cos\theta)$ (see Figure 1.5 (left-hand side)). We write:

$$\begin{aligned}\mathcal{R}f_\theta(t) &= \int_{M(x,y)\in D_t} f(M)dM \\ &= \int_{-\infty}^{\infty}\int_{-\infty}^{\infty} f(x,y)\,\underbrace{\delta(t - x\cos\theta - y\sin\theta)}_{0 \text{ if } M\notin D_t \text{ else } 1}\,dx\,dy\end{aligned} \qquad [1.4]$$

where t is the coordinate along the direction θ and δ is the Kronecker symbol.

We have seen (section 1.2.2) that the transmitted intensity through the object recorded by a detector $\ln(I_0/I)$ is the integral of $\mu(x, y)$ along the line going through the object. In other words, an X-ray radiograph can be considered as the Radon transform of the linear absorption coefficient of the illuminated specimen.

Having recorded projections of the objects at enough θ angles, the reconstruction of the initial object can then be obtained on the basis of the Fourier slice theorem (see Figure 1.5). Indeed, the Fourier transform of a particular projection $\mathcal{R}f_\theta(t)$ is equal to a slice through the Fourier transform of $f(x, y)$.

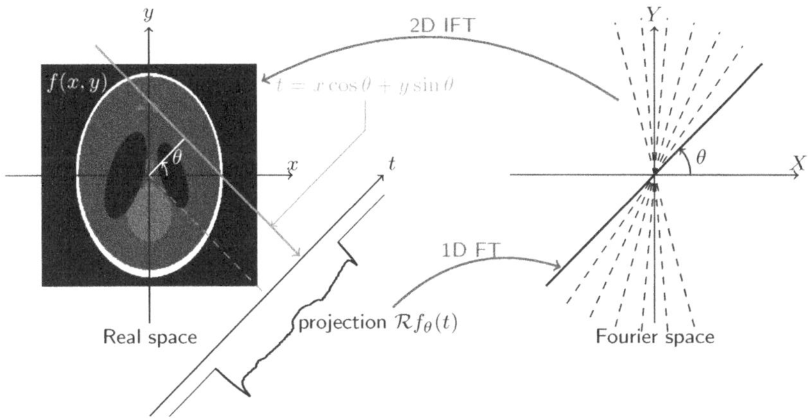

Figure 1.5. *Radon transform in parallel geometry of an object $f(x, y)$ (here the Shepp–Logan phantom) illustrating the Fourier slice theorem*

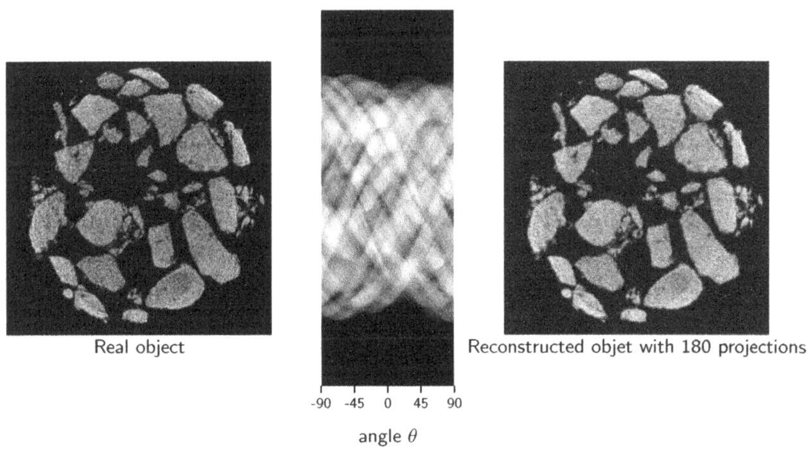

Figure 1.6. *Example of simulated reconstruction of small AgSn particles, using 180 projections and a Ram-Lak filter*

Figure 1.6 shows an example of the classic Ram-Lak filtered back projection reconstruction using 180 projections of a real object (small AgSn particles). The sinogram (assembly of all projections) is visible in the middle, and the reconstructed image agrees well with the initial object. A more precise look reveals radial streaks, which are typically due to under-sampling in the Fourier space on moving away from the rotation axis, and can be reduced by increasing the number of projections.

1.2.5. *Synchrotron X-ray microtomography*

Synchrotron radiation (SR) refers to the radiation emitted by relativistic charged particles (i.e. electrons most of the time) deviated by a magnetic field. Being observed in a small storage ring used for particle physics in 1947, it was first considered a problem because the accelerated particles were radiating and thus loosing energy. Soon researchers realized that they could actually build a synchrotron to serve as an intense X-ray source in their experiments [POL 83].

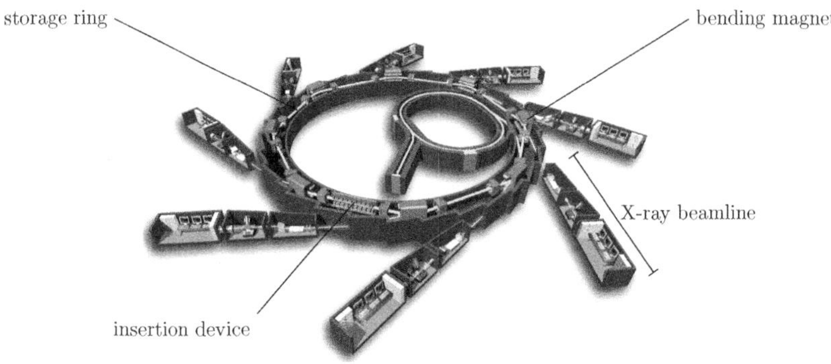

Figure 1.7. *Principle of synchrotron radiation emission from the storage ring (image from Soleil synchrotron)*

Since then, many synchrotrons have been built worldwide on the same principle: the electrons are traveling in a storage ring composed of a succession of straight lines and bending magnets to achieve a closed loop (Figure 1.7). This storage ring runs under a very high vacuum to minimize any possible collision between travelling electrons and residual gas molecules. The synchrotron light is produced by either the bending magnets or insertion devices (wigglers or undulators) installed on the straight line sections. Energy lost by the radiating electrons is restored thanks to radio frequency cavities installed all over the storage ring. Experimental beamlines are installed further below each bending magnet or insertion device to use the produced X-rays for actual experiments (see Figure 1.7). X-ray beams are usually characterized by their *brilliance* expressed in photons/s divided by the beam collimation in mrad (both horizontal and vertical), divided by the source area in mm^2 and for a 0.1% energy bandwidth of the considered energy. The brilliance of synchrotron radiation is more than 10 orders of magnitude higher than that of a typical laboratory X-ray source. An immediate consequence of this is that it is possible to work with a very narrow bandwidth of the X-ray spectrum (for instance, tuning a double-crystal X-ray monochromator) still having enough photons to carry out fast experiments.

X-ray synchrotron tomography typically uses an extended monochromatic parallel beam. The sample sits on a rotation stage and is illuminated by the beam to record radiographs on a detector. Many radiographs are recorded for each rotation angle to construct the sinogram. Thanks to the parallel geometry, the recorded projections are the Radon transform of the object for each particular rotation angle. The data set can then be processed by filtered back projection to produce a three-dimensional quantitative map of the linear attenuation coefficient $\mu(x, y, z)$. Because μ depends only on the density and the local average atomic number, the 3D image obtained usually reveals much information about the sample. Multiple phases can be distinguished provided their average atomic number is different and damages such as voids and cracks are visible as they create high-density variations.

In parallel geometry, the spatial resolution is controlled by the detector and may impose limits on the specimen size (in the horizontal plane) to observe. Typical microtomography installations make use of a scintillator + optics + high-throughput 2048×2048 pixels CCD to form the radiographs (see section 1.2.3). For a spatial resolution of 0.7 μm, the field of view in millimeters is then given by $FOV = 2048 \times 0.0007 = 1.4$ mm. Current setups readily achieve 0.7 μm effective spatial resolution and may be tuned (using thinner scintillator and a magnifying optics) to $\sim$ 0.2 μm. Below this value, this setup faces several technological limits and it is better to use magnified nano-tomography [MOK 07]. A variety of examples using different resolutions are shown in Figure 1.9.

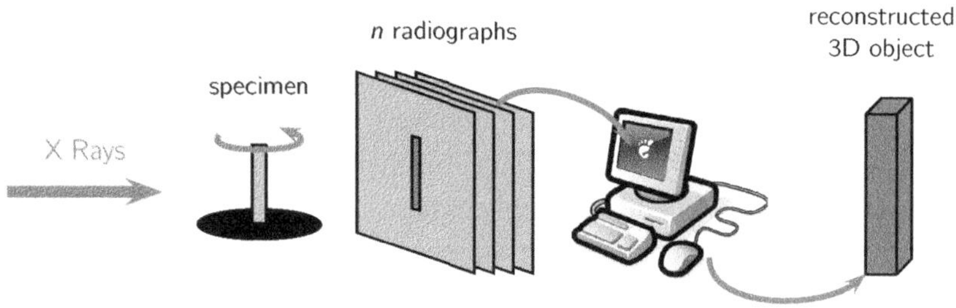

Figure 1.8. *Principle of synchrotron X-ray tomography*

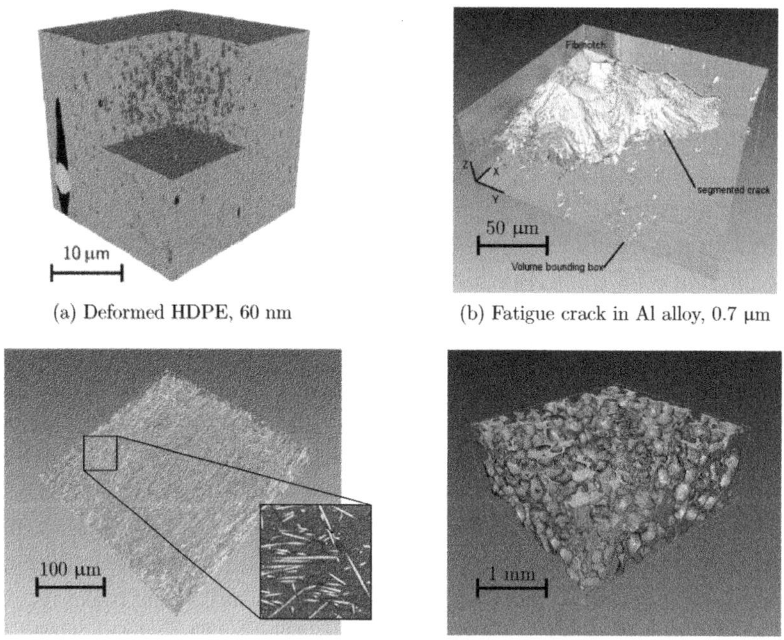

(a) Deformed HDPE, 60 nm

(b) Fatigue crack in Al alloy, 0.7 µm

(c) PA66 + glass fibers composite, 1.3 µm

(d) Ceramic foam, 10 µm

Figure 1.9. *Example of various objects and material images by X-ray tomography, the spatial resolution is indicated for each case. For a color version of the figure, see www.iste.co.uk/brancherie/microstructure.zip*

1.2.6. *Phase contrast tomography*

When an X-ray beam passes through a sample, it is not only partly absorbed as already described but also refracted. The refractive index is extremely close to 1, so it is usual to think of X-ray propagating without being deflected. Nevertheless, the refractive index of the material can be written as:

$$n = 1 - \delta + i\beta \qquad [1.5]$$

where β is the change in the amplitude of the wave, which is thus related to the absorption coefficient as $\beta = \lambda\mu/(4\pi)$, and $(1 - \delta)$ is the change in the phase of the wave. δ is of the order of 10^{-5} and much higher than β, which means that the phase contrast is more sensitive than absorption contrast. Imaging δ instead of β can therefore be of particular interest for low-absorbing materials, such as biological samples, and many interferometric-based techniques have been developed for this purpose [ALS 11].

A high degree of lateral coherence of synchrotron radiation allows another route, called in-line phase contrast imaging, to exploit this additional contrast mechanism.

Here, we consider a plane wave propagating in the x direction and illuminating a heterogeneous sample, characterized by $n(x, y, z)$, before being collected by a detector perpendicular to x placed at a distance D from the sample. It can be shown that the transmitted intensity through the sample I_D recorded by the detector can be written [CLO 97] as:

$$\ln \frac{I_0}{I_D} = \int \mu(x, y, z)dx + D \underbrace{\left(\frac{\partial^2}{\partial x^2} + \frac{\partial^2}{\partial y^2}\right)\left(\int n(x, y, z)dx\right)}_{\text{Laplacian of the projection of } n(x,y,z)} \qquad [1.6]$$

The reconstruction of the second term provides the 3D Laplacian of the real part of the refractive index $n(x, y, z)$; thus, the reconstructed quantity $o(x, y, z)$ from the projection images (classic FBP algorithm) is a good approximation:

$$o(x, y, z) = \mu(x, y, z) + D\Delta n(x, y, z) \qquad [1.7]$$

If $D = 0$ (detector place right behind the sample), then there is no phase contrast and $\mu(x, y, z)$ is obtained. If $D > 0$ but remains small (Fresnel regime), then the reconstructed image is the sum of two terms: the linear attenuation coefficient distribution μ and a second term linked to the Laplacian of n. In practice, interfaces and voids within a material are a source of discontinuity of n that will be picked up by Δn. This will produce black/white fringes at those particular locations resulting in an edge detection mode, which can be particularly useful to inspect small cracks in structural materials, for instance [BUF 99].

1.2.7. *Diffraction contrast tomography*

DCT is a monochromatic X-ray beam 3D imaging technique combining the principles of 3D X-ray diffraction microscopy [POU 04] and X-ray microtomography. During a $360°$ continuous rotation of the sample, each grain runs through a series of diffraction alignments, giving rise to diffracted beams. Part of these diffraction spots are captured on the high-resolution imaging detector system, positioned closely behind the sample [LUD 08, JOH 08]. Like in conventional micro-tomography, we can determine the three-dimensional distribution of the X-ray attenuation coefficient from the attenuation in the transmitted beam (see section 1.2.4), which is typically used to provide a mask of the sample geometry.

The analysis of the diffracted beams provides access to the crystalline microstructure. The processing route is described in [LUD 09b, REI 13]. After segmentation, Friedel pairs of diffraction spots (hkl and -h-k-l reflection from the same grain) are automatically identified ("pair matching"). A polycrystal indexing algorithm based on the analysis of such pairs of diffraction spots classifies the diffraction spots according to their grain of origin and determines the average orientation and elastic strain state of the grains (see

Figure 1.10). The 3D shape of each grain identified in the volume is obtained by algebraic iterative reconstruction [VAN 16] using the segmented diffraction spots as parallel projections of the unknown 3D grain volume at each recorded rotation position.

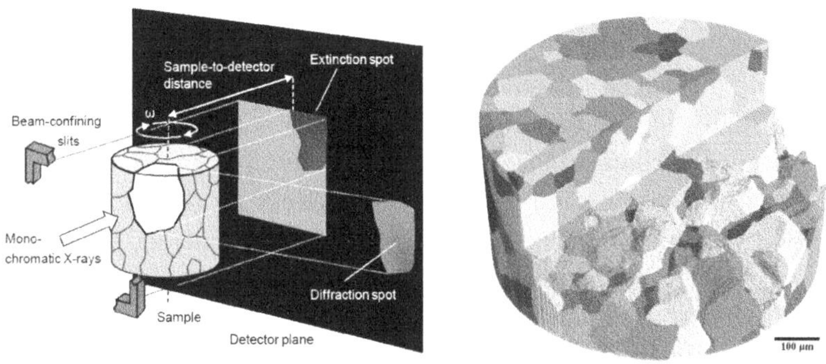

(a) (b)

Figure 1.10. *Illustration of the DCT technique (a) acquisition geometry, (b) example of reconstructed single-phase polycrystalline microstructure [LUD 09b, REI 13]. For a color version of the figure, see www.iste.co.uk/brancherie/microstructure.zip*

Using an extended box beam to illuminate the specimen allows to reconstruct the entire microstructure with a single scan provided the grains have a low orientation spread (<1° typically) and the number of grains within the shined volume do not produce significant overlap of the diffraction spots [LUD 10]. The typical acquisition time is on the order of an hour on a high-brilliance beam line such as ID11 at ESRF (the time is highly material dependent), which makes it the fastest X-ray grain mapping technique.

Recently, iterative reconstruction frameworks based on forward simulation have emerged and greatly increased the capability and precision of the technique [VIG 14].

1.3. Nanox: a miniature mechanical stress rig designed for near-field X-ray diffraction imaging techniques

Diffraction contrast tomography (DCT) allows to retrieve the three-dimensional microstructure of an undeformed polycrystalline sample (see section 1.2.7). Allowing a sufficient angular range to capture several diffraction spot families is a geometric requirement requiring the detector to stand only a few millimeters from the sample. This usually means that for combining this sort of characterization with actual mechanical loading, we would do a prior DCT experiment and then mount the sample in the actual stress rig (see, for instance, [HER 11]). This prevented to carry out the DCT *in situ* and also caused several issues with image registration between prior DCT and subsequent tomographic analysis using absorption/phase contrast, for instance. All of this made a mechanical loading machine compatible with DCT acquisition geometry highly desirable.

Figure 1.11 shows a schematic of a mechanical stress rig developed in collaboration between MINES ParisTech and INSA Lyon within the ANR CRYSTAL project [GUE 16]. The requirements were as follows:

– a tension load up to 500 N;

– a cyclic load at several 10 Hz;

– a load measurement better than 0.5 N;

– the observation point of the specimen must be located at the ID11 diffraction center when mounted at a height of 65 mm from the base plate;

– a ID11 Frelon detector with revolving optics may approach the rotation center as close as possible (< 3 mm);

– a 360° rotation for sub-micron tomographic acquisition.

The strong space constraints were fulfilled using a modified piezoelectric actuator (from DSM, USA). With the present design, a FReLon camera with a revolving optic system can be as close as 2.5 mm to the center of rotation (0.5 mm from the glass tube) and allow for 360° rotation to perform high-quality PCT and/or DCT scans. No existing load cell for tension testing could fit in this design, so the loading axis was instrumented with a full Wheatstone bridge of semiconductor strain gages (Texens, France). The strain gage signal is amplified and conditioned (tunable gain) to lead a 0–10 V signal over 12 bits.

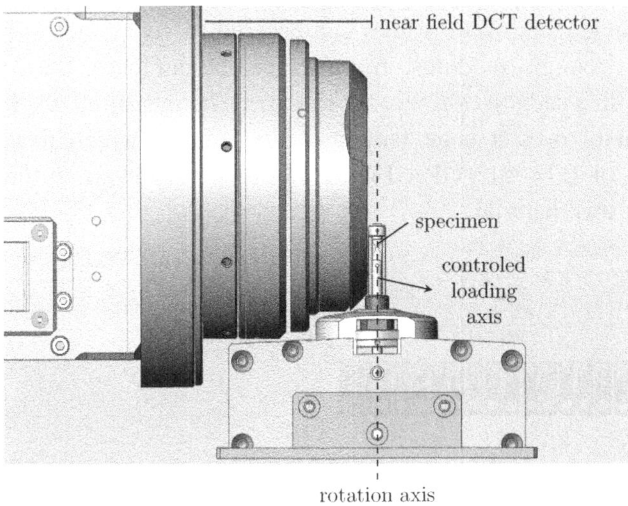

Figure 1.11. *Nanox a new* in situ *tension/fatigue stress rig for DCT [GUE 16]*

The piezoelectric actuator runs on a voltage range of $-30/+150$ V, which results in a vertical $-16/+80$ μm motion. A software-controllable voltage generator capable of supplying alternative sinusoidal waveforms for cyclic loading is used. All cables ideally pass through the slip rings of the diffractometer (like that at ID11) so that they have no impact on the rotation quality for tomographic acquisition. Integration within the beamline control and acquisition software is also essential to conduct fully automated mechanical testing and imaging. This is key to progress toward high-throughput 4D mechanical testing.

It is useful to measure the macroscopic strain applied to the sample. Regarding the tiny design of the device, a virtual extensometer consisting of two Pb balls glued on the sample surface has been tested (see Figure 1.12(a)). Tracking the displacement of the balls in the direct beam allows to know how much the sample has been elongated. The X-ray radiographs are automatically processed by a Python script: the two balls are labeled and the center of mass is computed (Figure 1.12(b)). The test was performed with a 316LN steel sample mounted on the device, and those values can be used to verify the calibration / measure Young's modulus, for instance, in our case, the load was 100 N with a section of 0.45×0.45 mm^2. The measured displacement is 1.8 pixel over a gage length of 769 pixels, which results in a modulus of E = 210 GPa (Figure 1.12(c)), very close to the known value for this material.

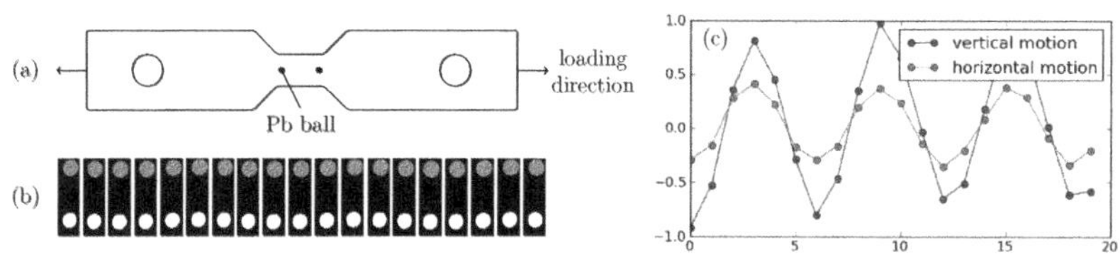

Figure 1.12. *X-ray virtual extensometer: (a) a small 316L tomographic sample with two 50 μm lead ball glued at each end of the gage length, (b) 100 N cyclic load is applied at 0.1 Hz, X-ray radiographs are recorded every 0.2 s and processed automatically, (c) displacement measured in micron*

The stress rig is also compatible with topotomography measurement [LUD 07], which is sensitive to crystal defects in the form of local variation of the crystal orientation or lattice spacing. For a particular (hkl) plane of a given grain, by aligning the normal of the lattice plane with the rotation axis, many X-ray topographs can be recorded and 3D topotomographic reconstruction can be achieved. Combined with *in situ* mechanical testing, this is a unique way to study plasticity and intra-granular damage within individual grains deeply embedded in the bulk of a metallic specimen.

The Nanox device was tested successfully for DCT imaging with a binary Al–Li alloy, which have been recrystallized to reach undeformed multi-crystalline specimen. A DCT scan of the specimen was conducted, and the (111) lattice plane of one of the grains in the gage length was aligned in topotomography. Figure 1.13 shows a summary of the study conducted at ID11, ESRF [GUE 16].

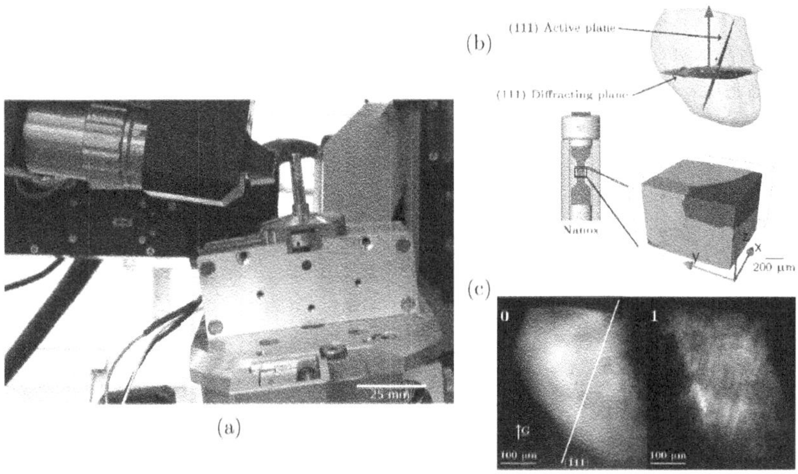

Figure 1.13. *DCT validation: (a) the Nanox stress rig mounted on the ID11 diffractometer for topotomography inspection, (b) the reconstructed gage section with the grain aligned showing the lattice planes of interest, (c) recorded X-ray Bragg topographs showing the band contrast appearing upon loading and localized along the* $(1\bar{1}1)$ *lattice planes of that grain*

1.4. Coupling diffraction contrast tomography with the finite-element method

1.4.1. *Motivation for image-based mechanical computations*

In recent years, continuum crystal plasticity has proved to be a powerful tool to interpret experimental results obtained in the deformation of metallic polycrystals [TEO 97]. Large-scale simulations on polycrystalline volume elements can now be performed with sufficient local discretization to predict the transgranular plastic strain fields. They are necessary for comparison with results of sophisticated 2D and 3D full-field measurements providing strain fields on the basis of grid methods and/or image correlation techniques [HER 07], lattice rotation fields by means of electron back-scattered diffraction (EBSD) [MAU 08, BRI 13a, BRI 13b] and elastic strain fields using, for instance, micro-diffraction techniques [ROB 11].

Even though continuum crystal plasticity constitutive equations do not account for the intricate dislocation mechanisms at work during plastic deformation, the continuum mechanical model gives relevant information about the strain heterogeneities induced in a polycrystal by the strong incompatibilities that develop at grain boundaries. These predictions are based on the precise kinematic description of the crystallographic nature of plastic slip in grains. For that purpose, full 3D simulations are necessary even when only 2D measurement information is available, due to the 3D nature of multi-slip mechanisms. Some of the first illustrations of the tremendous heterogeneities that develop in polycrystals are provided in [BAR 01a, BAR 01b, ZEG 07].

As stated above, synchrotron X-ray diffraction using fast 2D imaging detectors such as diffraction contrast tomography (DCT) and high-energy diffraction microscopy (HEDM) can now provide complete three-dimensional reconstruction of the grain microstructure of millimeter-sized metallic samples with several thousand grains. Forward modeling methods can further be used to extract local intragranular lattice rotations in each of the reconstructed grain [SUT 06, VIG 14]. This will probably be one of the key features used in the future to analyze material deformation mechanisms and degradation. Serial sectioning techniques are also rapidly developing, such as FIB/EBSD imaging and femtosecond laser ablation techniques coupled to EBSD (TriBeam experiments) [ECH 12, ECH 15].

All these new experimental developments can provide 3D microstructure images to serve as input for finite-element calculations of the local/global mechanical properties.

1.4.2. *3D mesh generation from tomographic images*

Microstructural description of polycrystalline (or mutli-phase) materials are generally provided by digital images (could be 2D or 3D) composed of patches of constant positive integers called labels and representing the different phases. When meshing a multi-labeled image, two radically different approaches are generally considered: (i) voxel-based meshing using hexahedra and (ii) free meshing using tetrahedra. Free meshing using hexahedra also exists and may play an important role in the future due to the better numerical precision achieved by those elements in comparison to tetrahedra, but algorithms are still under development [KRE 13].

Choosing between meshing methods usually depends on the goal to achieve, but it can be observed that when working in 3D and depending of the spatial resolution, voxel-based meshing may lead to prohibitive computational cost. Let us take a cubic grain of size $d = 50$ μm. For voxel-based meshing, the number of elements can readily be calculated as $(d/a)^3$, with a the element size at the grain boundary. For free meshing with tetrahedron, a geometric progression of the element size can be used with coarser elements at the center of the grain, for this value of q. For a progression $q > 1$, the total number of elements can be approximated by $6(d/a)^2 \times 2n$, with n being the number of elements from the grain boundary to the middle of the grain. The factor 6 comes from the regular division of a cube into six tetrahedra, and n is thus linked to half the grain size by $n = a + qa + \cdots + q^{n-1} = a(1-q^n)/(1-q) = d/2$ and can be calculated by $n = \log(1+(q-1)$ $d/2a)/\log(q) \propto \log(d/a)$. Finally, the total number of elements scales in $(d/a)^2 \log(d/a)$ instead of $(d/a)^3$. Table 1.1 compares the number of elements required to mesh our 50 μm grain with a progression of $q = 2$ and different spatial resolutions.

a (μm)	Voxel-based meshing	Free meshing
1.0	125 k	150 k
0.5	1 000 k	720 k
0.2	15 625 k	5 250 k
0.1	125 000 k	24 000 k

Table 1.1. *Comparison of the number of elements required to mesh a 50 μm grain with an element size equal to a at the grain boundary, between voxel-based and free meshing with a progression of $q = 2$*

If we relax the constraint of having a constant spatial resolution at the grain boundary but only at the edges of the grain boundaries (triple lines), the comparison is even more in favor of free meshing, because it now scales like $[\log(d/a)]^3$. Indeed, with the same kind of arguments, the element size is a at the edges of the grain boundary and increases to aq^{n-1} at the center of the grain boundary with $n \propto \log(d/a)$. Therefore, the number of elements on the grain boundary scales like $[\log(d/a)]^2$, and the total number of element in the grain scales like $[\log(d/a)]^3$.

Modern meshing software packages now allow to control the mesh density using a metric based on the proximity of features, and this should be typically used to produce a high-fidelity mesh of a multi-labeled image with the constraint of a reasonable number of elements (which can still reach several millions).

1.4.2.1. *Voxel-based meshing of multi-labeled images*

Voxel-based meshing consists of building a regular mesh of the size of the image and then grouping elements corresponding to each grain into element sets. The grain orientation can then be assigned to each particular set.

Two examples are presented here to illustrate this method. The first example is a Ni-based superalloy imaged by EBSD with a 1 μm step size (see Figure 1.14). The meshing was done directly as this resolution with an element size of 1 μm. After fitting a nonlinear crystal plasticity behavior to the macroscopic stress/strain curve, an average axial strain of 1% was applied using a generalized plan strain formulation to assess the heterogeneous stress distribution within the grains (see Figure 1.14(c)).

In the second example, a 316LN stainless steel sample imaged by EBSD via serial sectioning (courtesy Th. Ghidossi and L. Signor at Prime Institute, see [SIG 16] for more details) is meshed. The grain aggregate is fitted with a crystal plasticity behavior [GUI 13], and fatigue cycles are computed to try to reproduce the experimentally observed fatigue crack initiation locations (see Figure 1.15). The image data have been down-sampled by a factor of 4 in both x and y

directions to limit the number of elements below 1 million to be able to compute several fatigue cycles with a nonlinear crystal plasticity behavior in a few days.

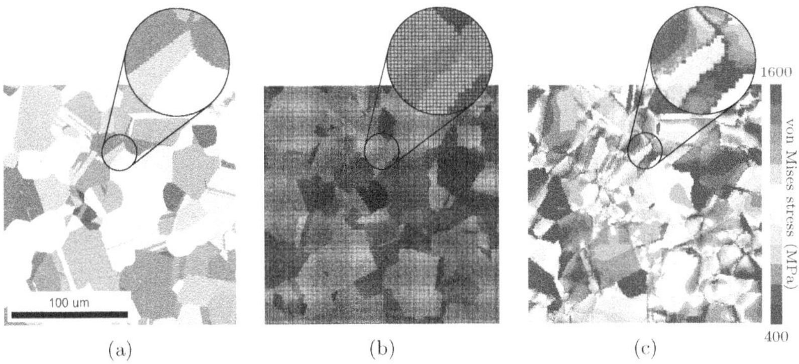

Figure 1.14. *(a) EBSD mapping of a Nickel-based superalloy sample (IPF coloring), (b) voxel-based meshing of the corresponding area (200 μm × 200 μm), (c) von Mises stress distribution after 1% axial strain. For a color version of the figure, see www.iste.co.uk/brancherie/microstructure.zip*

In summary, voxel-based meshing is fast and remains easy to implement. It has however two major drawbacks: it produces staircase-like grain boundaries due to the regular mesh and the number of elements scales very rapidly with the spatial resolution. This type of discretization is used for FFT-based mechanical calculations [ROL 10], and the argument holds true also in that case although the higher numerical efficiency of the algorithm intrinsically allows to target larger number of elements [BEL 11, LEB 08, LEB 12].

Voxel-based meshing may also be useful when considering grain microstructure with non-uniform orientation by grain. In this case, the orientation needs to be specified on a per-element basis, and a voxel-based mesh may be the more convenient way to achieve that, but to date this remains largely unexplored.

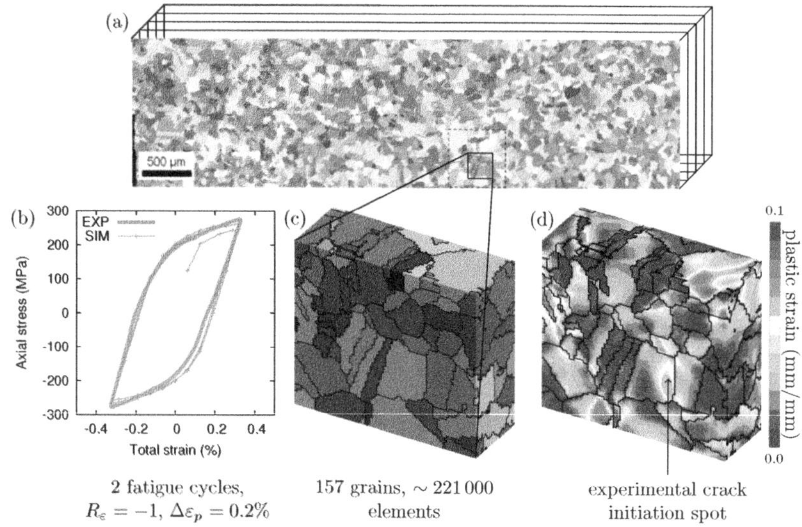

Figure 1.15. *(a) Serial sectioning EBSD mappings of a 316LN sample after 5,000 fatigue cycles, the total volume analyzed was* $5.0 \times 1.5 \times 0.127$ *mm*^{3} *with several 10 thousands of grains and 41 surface cracks (courtesy L. Signor, Prime Institute), (b) comparison of the stress–strain curve measured and simulated on the grain aggregate, (c) voxel-based 157 grain aggregate meshed from the EBSD data, (d) accumulated plastic strain after the second fatigue cycle showing some of the hot spots observed experimentally. For a color version of the figure, see www.iste.co.uk/brancherie/microstructure.zip*

1.4.2.2. *Free meshing of multi-labeled images*

The construction of a realistic mesh with tetrahedra finite elements from a multi-labeled tomographic image representing a polycrystalline microstructure is a non-trivial procedure.

In the simplest case, the microstructure image is composed of only one slice (surface EBSD, for instance). This is the case, for instance, of thin polycrystalline films, which exhibit a columnar grain structure in the film thickness direction [SIS 07]. The microstructure is then mesh in the plane and simply extruded to represent the underlying material. This methodology has been repeatedly used in the past to conduct crystal plasticity simulation of individual grain responses and compare it with surface measurements by digital image correlation [KAM 13, TAS 14, CHE 16, STI 16b] or high-resolution EBSD [BRI 13a, BRI 13b].

The general case of a real 3D microstructure is much more challenging. Recent progress with various 3D imaging techniques of polycrystalline microstructures triggered new software developments to handle such images. We can cite the open-source DREAM3D software developed by [GRO 14] or commercial suites like Avizo [AVI] or Simpleware [SIM].

New methods have been recently published and may greatly ease this process although not yet readily available. F. N'Guyen used an original approach purely based on morphological mathematics to mesh multi-phased structures without any ambiguity [N'GU 14]. With a completely different philosophy, level set functions coupled to anisotropic re-meshing can be used to obtain a high-fidelity mesh of any 3D image [ROU 11]. This method is currently applied on 3D images based on Laguerre tessellations to study recrystallization [BER 08, BER 11]. The last approach that is worth mentioning is the one carried out by R. Quey and implemented in the open-source software Neper [QUE 11]. Besides the usual voxel-based and Voronoï tessellation meshing, Neper can also be used to tweak a tessellation based on the center of mass of the grains to mimic the experimental microstructure.

Here, the meshing strategy consists of three main steps: surface mesh generation of the grain boundaries, simplification and filling the surface mesh with tetrahedral volume elements. Finally, a 3D FE mesh with a high fidelity with respect to the polycrystalline sample is obtained [PRO 16a]:

– *Surface mesh generation.* The 3D grain boundary surface mesh is typically generated by a multiple material marching cubes algorithm [WU 03] applied to a segmented image to generate a triangulated surface mesh. Subsequent surface smoothing is carried out by a Laplacian filter, which removes most of the staircase artifacts typical of marching cubes algorithms.

– *Mesh simplification.* Depending on the spatial resolution of the input 3D image, the surface mesh created at this point may contain millions of sub-voxel-sized elements. Mesh coarsening is thus carried out using an iterative decimation approach by collapsing edges into vertices in a given range, which allows to preserve mesh topology [GUE 97]. Minimum and maximum edge lengths are to be well chosen at this point because those values directly control the mesh density on the grain boundaries in the final volumic mesh.

– *Volume mesh generation.* Once the surface mesh of the grain boundaries is obtained, the volume mesh generator Ghs3d, developed by Inria and distributed by Distene [GHS], is used. The volume mesh generation is based on the Delaunay method described in detail in [GEO 97]. A tetrahedral mesh suitable for FE calculation and consistent with the geometry of the grain boundaries is created. Finally, a mesh optimization step is performed to improve the mesh quality.

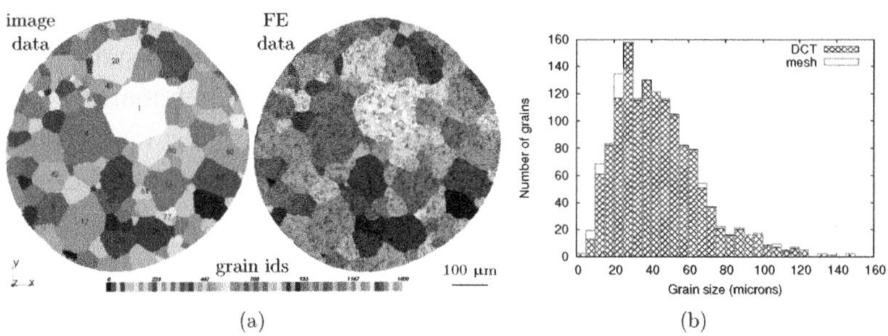

Figure 1.16. *(a) One horizontal slice of the tomographic images colored by grain number on the left and the corresponding slice of the FE mesh on the right; (b) comparison of 3D grain size distributions in the original DCT image and in the FE mesh (pure Ti sample). For a color version of the figure, see www.iste.co.uk/brancherie/microstructure.zip*

– *Relabelling grains.* The last step of the mesh generation consists of re-labeling the grains in order to have the same grain label as in the tomographic images (see Figure 1.16(a)), to assign reliably the measured grain orientation to each grain of the FE mesh. For this, the center of mass of each grain in the mesh is calculated from the center of mass of its elements. Then, the corresponding grain number is determined in the tomographic images and assigned to the FE mesh. This, of course, works only if the FE-produced mesh is of high fidelity and if the grain shapes are not too tortuous, but it proved to work remarkably well for the materials investigated here.

The geometric precision in terms of grain boundary positions of the FE mesh reconstructed by this method has been analyzed showing a very good agreement between the mesh and the tomographic image (see Figure 1.16(b)). Selection of meshes obtained by the present method is shown in Figure 1.17.

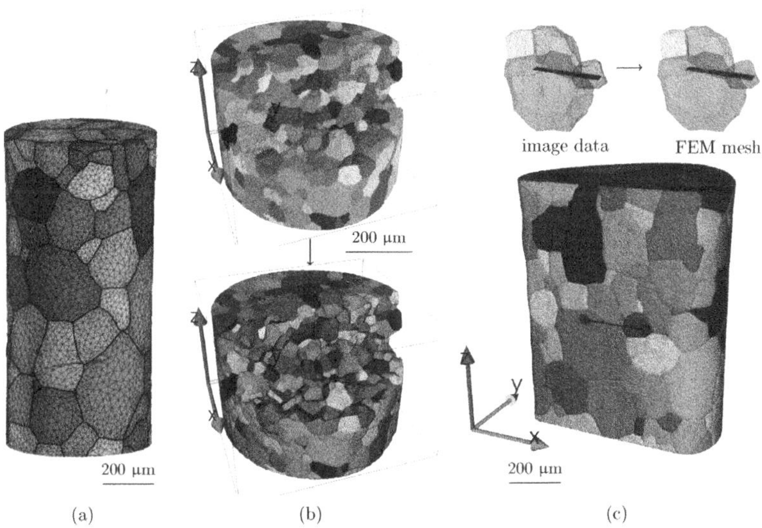

Figure 1.17. *(a) β-Ti 130 grain sample [LUD 09a], (b) both image data and FE mesh for the pure-Ti 1400 grains sample (some of the grains were removed to see the interior of the sample), (c) 55531 β-Ti sample with 387 grains and a central crack, a close up view of some of the grains of interest is shown for both image data and the corresponding FEM mesh. For a color version of the figure, see www.iste.co.uk/brancherie/microstructure.zip*

1.4.3. *Toward a fatigue model at the scale of the polycrystal*

Short fatigue crack propagation in the first few grains of a polycrystalline microstructure remains extremely challenging to model and predict, although in many cases it may represent a very significant portion of component life [CHR 14]. The effect of microstructural variability on initiation has been revived with the advent of 3D microstructure characterization tools paired with the possibility to carry out large-scale crystal plasticity simulations [CAS 14]. This has led to real advances in our understanding of fatigue crack initiation, as well as in our ability to predict it [MCD 10, SAN 11, STI 16a].

On the contrary, the growth behavior remains difficult to analyze because the stress state or the plastic activity at the crack tip for a given number of cycles remains unknown. One promising route seems thus to couple detailed 3D experiments with microstructure-sensitive CPFE computations *including the crack geometry* to account for the stress and plastic strain redistribution. Indeed, the general way to look at this problem is to compute some fatigue indicator parameters (FIP), which are to be regarded as crack formation driving forces [CAS 12]. FIPs have been used extensively for predicting the initiation phase [FAT 88, TSC 08, HOC 10, MUS 12, ROB 12] and are gaining traction for the stage I propagation phase analysis [SPE 14, CAS 16, SPE 16].

In spite of this, due to a lack of experimental comparisons, it is still not clear if predictions based on FIP distributions are accurate. One reason is the technical limitations to simultaneously account for the geometry of the crack, the microstructural features and a large enough volume of material to avoid side effects, as well as simulate many fatigue cycles.

A model described in [PRO 16b], previously used only with single crystal configurations using a different combination of slip systems in the mechanical behavior, has been used with an experimentally measured 3D microstructure to propagate a fatigue crack through several grains [PRO 17].

The three-dimensional grain structure was captured by DCT. The reconstructed data set is a 3D volume of the grain labels, in the laboratory coordinate system. Each label is complemented by the mean grain orientation from the diffraction indexing in the form of a Rodrigues vector expressed in the same coordinate system, which can be transformed into Euler angles for the simulation. The mesh includes the 387 grains imaged by DCT and the initial FIB notch used to initiate the fatigue crack (see Figure 1.17(c)). The complete calculation is 18 million dof, and several propagation steps are run using parallel resolution. The crack propagates mainly within three grains and depicts several features observed experimentally. On the left-hand side, the crack propagates mainly on a (110) plane highly activated and continues to do so after the first grain boundary, because another (110) plane is activated with very low twist angle at the next grain boundary. Due to this favorable geometrical configuration, this branch has the fastest crack growth. On the right-hand side, the crack adopts a zigzag pattern due to a less favorable configuration and is slower. This is in general agreement with the experiment although the details of the crack path are different. In the future, detailed experimental and simulated crack path comparisons will be essential to move toward a comprehensive modeling of the short crack behavior.

1.5. Conclusion and outlook

Determining microstructure–property relationships is directly linked to our ability to observe both the microstructure and the deformation/failure mechanisms concurrently. This field is moving rapidly with new means of observing the microstructure non-destructively in 3D (such as DCT, among others) and subsequent *in situ* imaging (SEM, DIC on the specimen surface or CT for internal damage).

Now is a particularly exciting time, because DCT characterization just became feasible using a dedicated laboratory X-ray setup [KIN 14, MCD 15]. The partnership between Zeiss and Xnovo (a spinoff from DTU) set in 2015 led to the commercialization of a DCT system on a laboratory X-ray microscope. Using a Laue focusing geometry, the labDCT setup provides the average orientation, center of mass and volume of all grains above 40 μm [MCD 15]. The upcoming version of the reconstruction software will also give access to the grain shapes (see the reconstructed AlLi microstructure in Figure 1.1(d)).

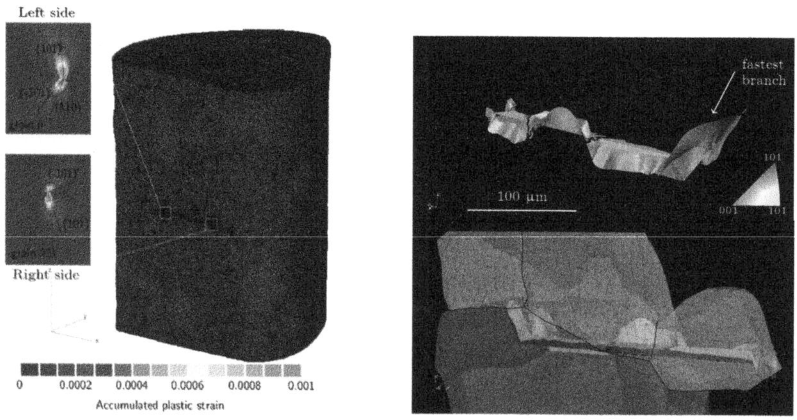

(a) (b)

Figure 1.18. *Short fatigue crack path prediction in a polycrystalline specimen from the DCT image: (a) initial configuration with the small FIB notch, (b) crack after four propagation steps showing a typical short crack behavior [PRO 17]. For a color version of the figure, see www.iste.co.uk/brancherie/microstructure.zip*

While DCT has become feasible in the laboratory, synchrotron X-ray imaging and diffraction continue to develop. Topotomography (TT) can be used to characterize crystal defects in 3D [LUD 07] and has recently allowed to observe the early stage of plasticity (typically below 1% strain) in bulk grains [GUE 16]. The development of a 6D-DCT framework now allows to measure quantitatively the local orientation field of the grains [VIG 14]. Hard X-ray microscopy (HXRM) is maturing and now allows orientation and strain mapping within a bulk grain with unparalleled precision [SIM 15, SIM 16]. Although HXRM is in practice difficult, measurements requiring long acquisition times, the recent development of a dedicated station at ID06 (ESRF) and the upcoming upgrade of the source (about 2 orders of magnitude in brilliance) will soon make it an unparalleled technique for bulk three-dimensional strain and orientation mapping.

1.6. Bibliography

[ALS 11] ALS-NIELSEN J., MCMORROW D., *Elements of Modern X-ray Physics*, 2nd edition, John Wiley & Sons, London, 2011.

[AVI] AVIZO, version 7, available at: http://www.fei.com/software/avizo-for-materials-science.

[BAR 01a] BARBE F., DECKER L., JEULIN D. *et al.*, "Intergranular and intragranular behavior of polycrystalline aggregates. Part 1: F.E. model", *International Journal of Plasticity*, vol. 17, no. 4, pp. 513–536, 2001.

[BAR 01b] BARBE F., FOREST S., CAILLETAUD G., "Intergranular and intragranular behavior of polycrystalline aggregates. Part 2: results", *International Journal of Plasticity*, vol. 17, no. 4, pp. 537–563, 2001.

[BEL 11] BELKHABBAZ A., BRENNER R., RUPIN N. *et al.*, "Prediction of the overall behavior of a 3D microstructure of austenitic steel by using FFT numerical scheme", *Procedia Engineering, 11th International Conference on the Mechanical Behavior of Materials (ICM11)*, vol. 10, pp. 1883–1888, 2011.

[BER 08] BERNACKI M., CHASTEL Y., COUPEZ T. *et al.*, "Level set framework for the numerical modelling of primary recrystallization in polycrystalline materials", *Scripta Materialia*, vol. 58, no. 12, pp. 1129–1132, 2008.

[BER 11] BERNACKI M., LOGÉ R.E., COUPEZ T., "Level set framework for the finite-element modelling of recrystallization and grain growth in polycrystalline materials", *Scripta Materialia*, vol. 64, no. 6, pp. 525–528, 2011.

[BRI 13a] BRITTON T.B., JIANG J., CLOUGH R. *et al.*, "Assessing the precision of strain measurements using electron backscatter diffraction – Part 1: detector assessment", *Ultramicroscopy*, vol. 135, pp. 126–135, 2013.

[BRI 13b] BRITTON T.B., JIANG J., CLOUGH R. *et al.*, "Assessing the precision of strain measurements using electron backscatter diffraction – Part 2: experimental demonstration", *Ultramicroscopy*, vol. 135, pp. 136–141, 2013.

[BUF 99] BUFFIÈRE J.-Y., MAIRE E., CLOETENS P. *et al.*, "Characterization of internal damage in a MMCp using X-ray synchrotron phase contrast microtomography", *Acta Materialia*, vol. 47, no. 5, pp. 1613–1625, 1999.

[BUF 10] BUFFIÈRE J.-Y., MAIRE E., ADRIEN J. *et al.*, "*In situ* experiments with X ray tomography: an attractive tool for experimental mechanics", *Experimental Mechanics*, vol. 50, no. 3, pp. 289–305, 2010.

[CAS 12] CASTELLUCCIO G.M., MCDOWELL D.L., "Assessment of small fatigue crack growth driving forces in single crystals with and without slip bands", *International Journal of Fracture*, vol. 176, no. 1, pp. 49–64, 2012.

[CAS 14] CASTELLUCCIO G.M., MUSINSKI W.D., MCDOWELL D.L., "Recent developments in assessing microstructure-sensitive early stage fatigue of polycrystals", *Current Opinion in Solid State and Materials Science*, vol. 18, no. 4, pp. 180–187, 2014.

[CAS 16] CASTELLUCCIO G.M., MCDOWELL D.L., "Microstructure-sensitive small fatigue crack growth assessment: Effect of strain ratio, multiaxial strain state, and geometric discontinuities", *International Journal of Fatigue*, vol. 82, Part 3, pp. 521–529, 2016.

[CHE 16] CHEN Z., DALY S.H., "Active slip system identification in polycrystalline metals by digital image correlation (DIC)", *Experimental Mechanics*, pp. 1–13, 2016.

[CHR 14] CHRIST H.-J., FRITZEN C.-P., KÖSTER P., "Micromechanical modeling of short fatigue cracks", *Current Opinion in Solid State and Materials Science*, vol. 18, no. 4, pp. 205–211, 2014.

[CLO 97] CLOETENS P., PATEYRON-SALOMÉ M., BUFFIÈRE J.-Y. *et al.*, "Observation of microstructure and damage in materials by phase sensitive radiography and tomography", *Journal of Applied Physics*, vol. 81, no. 9, pp. 5878–5886, May 1997.

[DUN 99] DUNN D.N., HULL R., "Reconstruction of three-dimensional chemistry and geometry using focused ion beam microscopy", *Applied Physics Letters*, vol. 75, pp. 3414–3416, 1999.

[ECH 12] ECHLIN M.P., MOTTURA A., TORBET C.J. *et al.*, "A new TriBeam system for three dimensional multimodal materials analysis", *Review of Scientific Instruments*, vol. 83, p. 023701, 2012.

[ECH 15] ECHLIN M.P., STRAW M., RANDOLPH S. *et al.*, "The TriBeam system: femtosecond laser ablation in situ SEM", *Materials Characterization*, vol. 100, pp. 1–12, 2015.

[FAT 88] FATEMI A., SOCIE D.F., "A critical plane approach to multiaxial fatigue damage including out-of-phase loading", *Fatigue & Fracture of Engineering Materials & Structures*, vol. 11, no. 3, pp. 149–165, 1988.

[GEO 97] GEORGE P.L., BOROUCHAKI H., *Triangulation de Delaunay et maillage: application aux éléments finis*, Hermes, Paris, 1997.

[GHS] GHS3D, available at: https://www.rocq.inria.fr/gamma/gamma/ghs3d, version 3.2-3.

[GRA 02] GRANGEAT P., *La tomographie. Fondements mathématiques, imagerie microscopique et imagerie industrielle*, Hermes-Lavoisier, 2002.

[GRO 14] GROEBER A., JACKSON A., "DREAM.3D: a Digital Representation Environment for the Analysis of Microstructure in 3D", *Integrating Materials and Manufacturing Innovation*, vol. 3, p. 5, 2014.

[GUE 97] GUEZIEC A., Surface simplification inside a tolerance volume, Report no. RC 20440, IBM Research Center, Yorktown Heights, New York, 1997.

[GUE 16] GUENINCHAULT N., PROUDHON H., LUDWIG W., "Nanox, a miniature mechanical stress rig designed for near-field X-ray diffraction imaging techniques", *Journal of Synchrotron Radiation*, vol. 23, no. 6, pp. 1474–1483, 2016.

[GUI 13] GUILHEM Y., BASSEVILLE S., CURTIT F. *et al.*, "Numerical investigations of the free surface effect in three-dimensional polycrystalline aggregates", *Computational Materials Science*, vol. 70, pp. 150–162, 2013.

[HER 07] HERIPRÉ E., DEXET M., CRÉPIN J. *et al.*, "Coupling between experimental measurements and polycrystal finite element calculations for micromechanical study of metallic materials", *International Journal of Plasticity*, vol. 23, no. 9, pp. 1512–1539, 2007.

[HER 11] HERBIG M., KING A., REISCHIG P. *et al.*, "3-D growth of a short fatigue crack within a polycrystalline microstructure studied using combined diffraction and phase-contrast X-ray tomography", *Acta Materialia*, vol. 59, no. 2, pp. 590–601, 2011.

[HOC 10] HOCHHALTER J.D., LITTLEWOOD D.J., CHRIST R.J. JR *et al.*, "A geometric approach to modeling microstructurally small fatigue crack formation: II. Physically based modeling of microstructure-dependent slip localization and actuation of the crack nucleation mechanism in AA 7075-T651", *Modelling and Simulation in Materials Science and Engineering*, vol. 18, no. 4, p. 045004, 2010.

[JIR 12] JIRUSE J., HRNCIR T., LOPOUR F. *et al.*, "Combined plasma FIB-SEM", *Microscopy and Microanalysis*, vol. 18, pp. 652–653, 2012.

[JOH 08] JOHNSON G., KING A., HONNICKE M.G. *et al.*, "X-ray diffraction contrast tomography: a novel technique for three-dimensional grain mapping of polycrystals. II. The combined case", *Journal of Applied Crystallography*, vol. 41, no. 2, pp. 310–318, April 2008.

[KAM 13] KAMMERS A.D., DALY S., "Digital image correlation under scanning electron microscopy: methodology and validation", *Experimental Mechanics*, vol. 53, no. 9, pp. 1743–1761, 2013.

[KIN 14] KING A., REISCHIG P., ADRIEN J. *et al.*, "Polychromatic diffraction contrast tomography", *Materials Characterization*, vol. 97, pp. 1–10, 2014.

[KRE 13] KREMER M., BOMMES D., LIM I. *et al.*, "Advanced automatic hexahedral mesh generation from surface quad meshes", *Proceedings of the 22nd International Meshing Roundtable*, Springer-Verlag, Berlin, 2013.

[LEB 08] LEBENSOHN R.A., BRENNER R., CASTELNAU O. *et al.*, "Orientation image-based micromechanical modelling of subgrain texture evolution in polycrystalline copper", *Acta Materialia*, vol. 56, no. 15, pp. 3914–3926, 2008.

[LEB 12] LEBENSOHN R.A., KANJARLA A.K., EISENLOHR P., "An elasto-viscoplastic formulation based on fast Fourier transforms for the prediction of micromechanical fields in polycrystalline materials", *International Journal of Plasticity*, vol. 32, pp. 59–69, 2012.

[LIE 11] LIENERT U., LI S.F., HEFFERAN C.M. *et al.*, "High-energy diffraction microscopy at the advanced photon source", *Journal of The Minerals, Metals & Materials Society (TMS)*, vol. 63, no. 7, pp. 70–77, 2011.

[LUD 07] LUDWIG W., LAURIDSEN E.M., SCHMIDT S. *et al.*, "High-resolution three-dimensional mapping of individual grains in polycrystals by topotomography", *Journal of Applied Crystallography*, vol. 40, no. 5, pp. 905–911, October 2007.

[LUD 08] LUDWIG W., SCHMIDT S., LAURIDSEN E.M. *et al.*, "X-ray diffraction contrast tomography: a novel technique for three-dimensional grain mapping of polycrystals. I. Direct beam case", *Journal of Applied Crystallography*, vol. 41, no. 2, pp. 302–309, April 2008.

[LUD 09a] LUDWIG W., KING A., REISCHIG P. *et al.*, "New opportunities for 3D materials science of polycrystalline materials at the micrometre lengthscale by combined use of X-ray diffraction and X-ray imaging", *Materials Science and Engineering: A*, vol. 524, nos. 1–2, pp. 69–76, 2009.

[LUD 09b] LUDWIG W., REISCHIG P., KING A. *et al.*, "Three-dimensional grain mapping by X-ray diffraction contrast tomography and the use of Friedel pairs in diffraction data analysis", *Review of Scientific Instruments*, vol. 80, no. 3, p. 033905, 2009.

[LUD 10] LUDWIG W., KING A., HERBIG M. *et al.*, "Characterization of polycrystalline materials using synchrotron X-ray imaging and diffraction techniques", *Journal of The Minerals, Metals & Materials Society (TMS)*, vol. 62, no. 12, pp. 22–28, Springer, December 2010.

[MAI 14] MAIRE E., WITHERS P., "Quantitative X-ray tomography", *International Materials Reviews*, vol. 59, no. 1, pp. 1–43, 2014.

[MAU 08] MAURICE C., FORTUNIER R., "A 3D Hough transform for indexing EBSD and Kossel patterns", *Journal of Microscopy*, vol. 230, no. 3, pp. 520–529, 2008.

[MCD 10] MCDOWELL D.L., DUNNE F.P.E., "Microstructure-sensitive computational modeling of fatigue crack formation", *International Journal of Fatigue*, vol. 32, no. 9, pp. 1521–1542, 2010.

[MCD 15] MCDONALD S.A., REISCHIG P., HOLZNER C. *et al.*, "Non-destructive mapping of grain orientations in 3D by laboratory X-ray microscopy", *Scientific Reports*, vol. 5, p. 14665, 2015.

[MOK 07] MOKSO R., CLOETENS P., MAIRE E. *et al.*, "Nanoscale zoom tomography with hard X-rays using Kirkpatrick-Baez optics", *Applied Physics Letters*, vol. 90, no. 14, p. 144104, April 2007.

[MUS 12] MUSINSKI W.D., MCDOWELL D.L., "Microstructure-sensitive probabilistic modeling of HCF crack initiation and early crack growth in Ni-base superalloy IN100 notched components", *International Journal of Fatigue*, vol. 37, pp. 41–53, 2012.

[N'GU 14] N'GUYEN F., Morphologie mathématique appliquée au développement d'outils de maillage EF automatique dans le cas de microstructures hétérogènes bi et multiphasées, PhD Thesis, University of Lille 1, 2014.

[PHI 09] PHILIPPE T., DE GEUSER F., DUGUAY S. *et al.*, "Clustering and nearest neighbour distances in atom-probe tomography", *Ultramicroscopy*, vol. 109, no. 10, pp. 1304–1309, 2009.

[POL 83] POLLOCK H.C., "The discovery of synchrotron radiation", *American Journal of Physics*, vol. 51, p. 278, 1983.

[POU 04] POULSEN H.F., *Three-dimensional X-ray diffraction microscopy – mapping polycrystals and their dynamics*, vol. 205 of *Springer Tracts in Modern Physics*, Springer, Berlin, 2004.

[PRO 16a] PROUDHON H., LI J., REISCHIG P. *et al.*, "Coupling diffraction contrast tomography with the finite element method", *Advanced Engineering Materials*, vol. 18, no. 6, pp. 903–912, 2016.

[PRO 16b] PROUDHON H., LI J., WANG F. *et al.*, "3D Simulation of short fatigue crack propagation by finite element crystal plasticity and remeshing", *International Journal of Fatigue*, vol. 82, Part 2, pp. 238–246, 2016.

[PRO 17] PROUDHON H., LI J., ROOS A. *et al.*, "Simulation of short fatigue crack propagation in a 3D experimental microstructure", *Advanced Engineering Materials*, p. 1600721, 2017.

[QUE 11] QUEY R., DAWSON P.R., BARBE F., "Large-scale 3D random polycrystals for the finite element method: generation, meshing and remeshing", *Computer Methods in Applied Mechanics and Engineering*, vol. 200, pp. 1729–1745, 2011.

[RAD 17] RADON J., "Uber die Bestimmung von Funktionen durch ihre Integralwerte langs gewisser Mannigfaltigkeiten", *Mathematisch-Physische Klasse*, vol. 69, pp. 262–277, 1917.

[REI 13] REISCHIG P., KING A., NERVO L. *et al.*, "Advances in X-ray diffraction contrast tomography: flexibility in the setup geometry and application to multiphase materials", *Journal of Applied Crystallography*, vol. 46, no. 2, pp. 297–311, April 2013.

[ROB 11] ROBACH O., MICHA J.-S., ULRICH O. *et al.*, "Full local elastic strain tensor from Laue microdiffraction: simultaneous Laue pattern and spot energy measurement", *Journal of Applied Crystallography*, vol. 44, no. 4, pp. 688–696, August 2011.

[ROB 12] ROBERT C., SAINTIER N., PALIN-LUC T. *et al.*, "Micro-mechanical modelling of high cycle fatigue behaviour of metals under multiaxial loads", *Mechanics of Materials*, vol. 55, pp. 112–129, 2012.

[ROL 10] ROLLETT A.D., LEBENSOHN R.A., GROEBER M. *et al.*, "Stress hot spots in viscoplastic deformation of polycrystals", *Modelling and Simulation in Materials Science and Engineering*, vol. 18, no. 7, p. 074005, 2010.

[ROU 11] ROUX E., BERNACKI M., BOUCHARD P., "A level-set and anisotropic adaptive remeshing strategy for the modeling of void growth under large plastic strain", *Computational Materials Science*, vol. 68, pp. 32–46, 2011.

[ROW 10] ROWENHORST D., LEWIS A., SPANOS G., "Three-dimensional analysis of grain topology and interface curvature in a [beta]-titanium alloy", *Acta Materialia*, vol. 58, no. 16, pp. 5511–5519, 2010.

[SAN 11] SANGID M.D., MAIER H.J., SEHITOGLU H., "An energy-based microstructure model to account for fatigue scatter in polycrystals", *Journal of the Mechanics and Physics of Solids*, vol. 59, no. 3, pp. 595–609, 2011.

[SEL 93] SELTZER S., "Calculation of photon mass energy-transfer and mass energy-absorption coefficients", *Radiation Research*, vol. 136, pp. 147–170, 1993.

[SIG 16] SIGNOR L., VILLECHAISE P., GHIDOSSI T. *et al.*, "Influence of local crystallographic configuration on microcrack initiation in fatigued 316LN stainless steel: experiments and crystal plasticity finite elements simulations", *Material Science and Engineering A*, vol. 649, no. 1, pp. 239–249, 2016.

[SIM] SIMPLEWARE, available at: http://www.simpleware.com.

[SIM 15] SIMONS H., KING A., LUDWIG W. *et al.*, "Dark-field X-ray microscopy for multiscale structural characterization", *Nature Communications*, vol. 6, Nature Publishing Group, 2015.

[SIM 16] SIMONS H., JAKOBSEN A.C., AHL S.R. *et al.*, "Multiscale 3D characterization with dark-field X-ray microscopy", *MRS Bulletin*, vol. 41, no. 6, pp. 454–459, Cambridge University Press, 2016.

[SIS 07] SISKA F., FOREST S., GUMBSCH P. *et al.*, "Finite element simulations of the cyclic elastoplastic behaviour of copper thin films", *Modelling and Simulation in Materials Science and Engineering*, vol. 15, no. 1, pp. S217–S238, 2007.

[SPE 14] SPEAR A.D., LI S.F., LIND J.F. *et al.*, "Three-dimensional characterization of microstructurally small fatigue-crack evolution using quantitative fractography combined with post-mortem X-ray tomography and high-energy X-ray diffraction microscopy", *Acta Materialia*, vol. 76, pp. 413–424, 2014.

[SPE 16] SPEAR A.D., HOCHHALTER J.D., CERRONE A.R. *et al.*, "A method to generate conformal finite-element meshes from 3D measurements of microstructurally small fatigue-crack propagation", *Fatigue & Fracture of Engineering Materials & Structures*, vol. 39, no. 6, pp. 737–751, 2016.

[STI 16a] STINVILLE J.-C., LENTHE W.C., MIAO J. *et al.*, "A combined grain scale elastic-plastic criterion for identification of fatigue crack initiation sites in a twin containing polycrystalline nickel-base superalloy", *Acta Materialia*, vol. 103, pp. 461–473, 2016.

[STI 16b] STINVILLE J.-C., ECHLIN M., TEXIER D. *et al.*, "Sub-grain scale digital image correlation by electron microscopy for polycrystalline materials during elastic and plastic deformation", *Experimental Mechanics*, vol. 56, no. 2, pp. 197–216, 2016.

[SUT 06] SUTER R.M., HENNESSY D., XIAO C. *et al.*, "Forward modeling method for microstructure reconstruction using X-ray diffraction microscopy: single-crystal verification", *Review of Scientific Instruments*, vol. 77, no. 12, p. 123905, 2006.

[TAS 14] TASAN C., HOEFNAGELS J., DIEHL M. *et al.*, "Strain localization and damage in dual phase steels investigated by coupled in-situ deformation experiments and crystal plasticity simulations", *International Journal of Plasticity*, vol. 63, pp. 198–210, 2014.

[TEO 97] TEODOSIU C., *Large plastic deformation of crystalline aggregates*, CISM Courses and Lectures No. 376, Udine, Springer Verlag, Berlin, 1997.

[TOD 13] TODA H., OHKAWA Y., KAMIKO T. *et al.*, "Grain boundary tracking: a four-dimensional visualization technique for determining grain boundary geometry via local strain mapping", *Acta Materialia*, vol. 61, no. 14, pp. 5535–5548, 2013.

[TSC 08] TSCHOPP M.A., MCDOWELL D.L., "Influence of single crystal orientation on homogeneous dislocation nucleation under uniaxial loading", *Journal of the Mechanics and Physics of Solids*, vol. 56, no. 5, pp. 1806–1830, 2008.

[UCH 12] UCHIC M., GROEBER M., SPOWART J. *et al.*, "An automated multi-modal serial sectioning system for characterization of grain-scale microstructures in engineering materials (preprint)", *AFNL Rep.*, pp. 195–202, 2012.

[VAN 16] VAN AARLE W., PALENSTIJN W.J., CANT J. *et al.*, "Fast and Flexible X-ray Tomography Using the ASTRA Toolbox", *Optics Express*, vol. 24, no. 22, pp. 25129–25147, 2016, http://dx.doi.org/10.1364/OE.24.025129.

[VIG 14] VIGANÒ N., LUDWIG W., BATENBURG K.J., "Reconstruction of local orientation in grains using a discrete representation of orientation space", *Journal of Applied Crystallography*, vol. 47, no. 6, pp. 1826–1840, December 2014.

[WEY 04] WEYLAND M., MIDGLEY P.A., "Electron tomography", *Materials Today*, vol. 7, no. 12, pp. 32–40, 2004.

[WU 03] WU Z. SULLIVAN J.M. JR, "Multiple material marching cubes algorithm", *International Journal for Numerical Methods in Engineering*, vol. 58, no. 2, pp. 189–207, 2003.

[ZEG 07] ZEGHADI A., FOREST S., GOURGUES A.-F. *et al.*, "Ensemble averaging stress–strain fields in polycrystalline aggregates with a constrained surface microstructure–Part 2: Crystal plasticity", *Philosophical Magazine*, vol. 87, no. 8-9, pp. 1425–1446, 2007.

2

Determining the Probability of Occurrence of Rarely Occurring Microstructural Configurations for Titanium Dwell Fatigue

Dwell fatigue failure is a critical concern for those who design disks made of near-α and α+β titanium alloys. Decades of prior work have demonstrated that microstructure, microtextured regions (MTRs) in particular, plays a key role in the overall susceptibility of a given material to dwell fatigue failure. Briefly, cracks are found to initiate in large hard-oriented MTRs that are neighbored by a large soft-oriented MTR due to a load-shedding mechanism. Dwell fatigue failure is a "weak link" event, and hence, it is necessary to know the distribution of hard and soft MTR sizes and the probability that they will neighbor one another within a given volume of material. As they measure multiple millimeters, it is impractical to characterize the full distribution of MTR sizes, especially the upper tail, using conventional methods like electron backscatter diffraction (EBSD). Hence, in this work, we propose two methods to identify the probability of occurrence of critical microstructure arrangements for titanium dwell fatigue. The first uses experimentally guided analysis of large volumes of synthetically generated microstructures, while the second is a hybrid of the first and the generalized extreme value theory (EVT).

Chapter written by Adam L. PILCHAK, Joseph C. TUCKER and Tyler J. WEIHING.

2.1. Introduction

Dwell fatigue failure of titanium alloys is a well-known phenomenon whereby a hold at peak stress can cause a significant reduction in the number of cycles to failure when compared with constant amplitude cyclic fatigue tests at equivalent conditions. The underlying mechanisms have been elucidated over the past few decades by experiments [WOO 95, PIL 11] and modeling [VEN 06, DUN 07, ANA 11]. The susceptibility of a given alloy to this deleterious mechanism is a strong function of alloy composition [QIU 14] and microstructure [WOO 96]. The near-α class of alloys is most susceptible and the vulnerability tends to decrease with the increasing β-stabilizer content. With regard to microstructure, features spanning a wide range of length scales can all contribute positively or negatively to dwell performance, but the presence and character of MTRs overwhelm all other factors. MTRs are local regions of microstructure in titanium alloys where the α phase (primary α and transformed β) shares a common c-axis orientation. In addition to the size and shape of the MTR, the work of Woodfield *et al.* [WOO 96] has demonstrated that the degree of crystallographic alignment within the MTR is another important factor.

Dwell fatigue failure is classically characterized by the occurrence of early subsurface crack nucleation [SIN 06] and rapid crack propagation across the initiating MTR by the faceted growth mechanism [SIN 06, PIL 11, PIL 13]. Sinha *et al.* [SIN 06] sectioned into one of these sites and found that the faceted region correlated with an underlying MTR with [0001] axes (c-axes) primarily aligned with the stress axis. Adjacent to this region were other MTRs with c-axes oriented ~90° to the stress axis. Elasticity theory [NYE 85] and single-crystal deformation experiments [WIL 02] have shown that these are the elastically and plastically hardest and softest orientations, respectively. The precise details of dwell fatigue crack nucleation and growth mechanism at atomic length scales are poorly understood currently, but it is known that room temperature creep enabled by dislocation glide in the soft MTR leads to local load shedding onto the hard MTR and these increased stresses lead to crack formation as demonstrated by experiment [EVA 94, SIN 06, PIL 11] and modeling

[VEN 06, DUN 07, ANA 11]. Once initiated, crack growth occurs rapidly through hard oriented α particles/grains with a relative insensitivity to crack length [PIL 11, PIL 13] and this rapid growth occurs throughout the entire initiating MTR. The crack growth rate is reduced to values encountered in conventional fatigue when the crack is growing in non-hard orientations. Thus, depending on the size of the first MTR to crack, a physically large crack can be present leading to severe debits in expected crack growth lifetime [PIL 14a].

The dwell fatigue response of a given lot of material shows a strong sample size effect [EVA 84, SON 88, SON 89] with larger test volumes exhibiting generally lower lifetimes, thus underscoring the "weak link" nature of this failure mechanism. Empirical data and results from modeling have shown that the worst microstructure occurs when large soft MTRs are adjacent to large hard MTRs for two reasons: (1) large volumes of materials deform by creep in the soft MTR, and hence, large stresses are shed onto the hard MTR, and (2) the large hard MTR will support rapid development of a physically large crack. Hence, the dwell susceptibility of a given lot of materials is increased when there is a high rate of occurrence of large hard MTRs neighboring large soft MTRs. A particularly bad case would be when two hard MTRs neighbor one another, producing a larger effective microstructural unit. Additionally, recall that the degree of crystallographic alignment is also an important factor, with MTRs exhibiting large volume fractions of closely aligned particles being more detrimental to dwell fatigue life. Due to the physical size of MTRs, which are typically measured in millimeters, it is difficult to fully characterize the size distribution of these features. First, large-area EBSD scans require specialized equipment and software, are experimentally challenging and are generally cost-prohibitive, at least for practical application. Second, once segmented by crystallographic orientation to define hard/soft MTRs and also screened according to crystallographic alignment, a given EBSD scan of 10 mm x 10 mm may yield a very small number of MTRs from which reliable metrics can be extracted. Hence, in this work, we propose a characterization-informed computational approach to provide an estimate of the number of critical microstructural arrangements in a volume of material. Physically large volumes of synthetic microstructure were generated in parallel using the Department of Defense (DoD) Supercomputing Resource Center (DSRC). These microstructures were screened for the occurrence rate of critical neighborhoods using criteria obtained by analyzing crack initiation sites on laboratory samples.

2.2. Experimental methods

The general procedure implemented in this work is as follows. Techniques for characterizing and quantifying microtexture are described. These statistics are used as inputs to synthetic microstructure building routines. The microstructural instantiations are screened using heuristic rules gleaned from experimental characterization. Scripting, batch processing and high-performance computing (HPC) enable us to create physically large volumes of materials that approach simulating a fleet of disks.

In this study, we perform the following calculations. First, we assess the variation in the density of critical MTRs as a function of MTR volume fraction under several simplifying assumptions. We treat the aggregate as a single-phase material with hexagonal crystal symmetry. Each grain in the structure represents an MTR. For these baseline investigations, we have modeled a completely random crystallographic distribution of MTRs with a quasi-unimodal size distribution. Then, we assess a combined synthetic building and EVT approach to quickly and efficiently make these assessments.

2.2.1. *MTR quantification metrics*

The first-order MTR parameters of interest to identify dwell fatigue susceptibility, and the propensity for microcrack initiation and propagation, are the size, density and c-axis misalignment. The size is an indicator of both how likely it is to nucleate a crack as a result of load shedding and also how far a nucleated crack will propagate before being arrested and/or slowed by encountering highly misoriented grains. Density is the fraction of pixels within an MTR that share a common c-axis orientation. It is defined as the fraction of "shrink-fit" pixels belonging to an MTR. The "shrink-fit" is the region searched, based on a multiple of grain average diameters, for common c-axis aligned grains. The fraction emerges from the grain pixels within the designated c-axis misalignment tolerance over the pixels checked. Pixels check is governed by the search range designated by the input multiple of grain average diameters. Density is a measure of how concentrated an MTR is with similarly aligned α grains. A higher density could be an indicator of worse acting regions than MTRs of similar size with lower densities. Finally, c-axis misalignment is a gauge for distribution of c-axes within MTRs. Highly aligned c-axes would imply the presence of low-angle grain boundaries between adjacent particles that provide limited resistance to crack progression. In our method, the c-axis misalignment of an MTR is taken as the average deviation of the c-axis of every pixel from the dominant c-axis orientation of the MTR.

As these features are interrelated with regard to the impact on mechanical properties, we define an intensity factor, which gives an indication of "badness" of a particular MTR in equation [2.1]:

$$I_{MTR} = \frac{\rho_{MTR} a_{MTR}}{\Delta g_{MTR}} \quad [2.1]$$

I_{MTR} is the MTR intensity, ρ_{MTR} is the MTR density, a_{MTR} is the MTR size and Δg_{MTR} is the c-axis misalignment. The intensity is a combined variable that physically represents the ratio of the normalized MTR density to the normalized MTR alignment. Density and size are directly proportional to intensity, as larger values indicate worse acting MTRs. C-axis misalignment is inversely proportional, in equation [2.1] since smaller values correspond to MTRs with a smaller fraction of high-angle boundaries, meaning that they are more likely to initiate and propagate microcracks. Ellipsoid (Ellipse)-fit semi-major axis could be used in place of size; this could be regarded as a more representative indicator of how far a nucleated crack will propagate. As correlations are made between these parameters and material behavior, we could envisage assigning weights to the quantities, as in equation [2.2], based on their relative influence.

$$I_{MTR} = \frac{w_{\rho} \rho_{MTR} w_{a} a_{MTR}}{w_{\Delta g} \Delta g_{MTR}} \quad [2.2]$$

2.2.2. Synthetic microstructure generation

In our methodology, synthetic microstructures were created using DREAM.3D [GRO 14]. The StatsGenerator Utility is used to input relevant microstructure statistics determined from the experiment. DREAM.3D uses an ellipsoid packing routine, the details of which are described elsewhere [TUC 12], to place grains in the simulated volume. In principle, DREAM.3D has the ability to produce very complex microstructures that closely approximate real titanium microstructures. For example, primary α particles phase size distribution, shape (aspect ratio) distribution, crystallographic orientation distribution function (ODF), morphological orientation distribution, misorientation distribution function (MDF) and neighbor distribution are all simultaneously optimized through an aggregate objective function. At the time of writing, synthetic building of the secondary α phase as discrete laths or colonies of α+β phase is not possible. The current best practice instead is to homogenize the secondary α phase into a single effective phase with the same aforementioned metrics as inputs. This does not affect this study since we are interested in multi-millimeter-length-scale features and hence we would not want the computational burden of storing discrete submicron laths on a voxelized grid. In fact, we make even further simplifications for our study as described in more detail below.

DREAM.3D filters are arranged in a pipeline, allowing the user modularity and extensibility in interpreting their data. In addition to the graphical user interface (GUI), DREAM.3D pipelines can be executed from the command line with an executable called "PipelineRunner". PipelineRunner takes a single argument pointing to the DREAM.3D pipeline file to be executed. A series of MATLAB routines called "AutoPipeLine" have been developed to batch run an arbitrary DREAM.3D pipeline using the command line capability. A design of experiments emerges through this function which we have used to first instantiate many synthetic structures with the same global statistics and, second, to systematically dissect those structures looking for critical microstructural neighborhoods. To this end, this batch processing capability was further extended and implemented on the DoD DSRC HPC Predator. Message passing interface (MPI) is used to send jobs to the different nodes on the HPC. Portable batch system (PBS) is used to schedule and checkout the requested nodes on Predator. This parallelization scheme is what is known as "embarrassingly parallel" in that each node performs exactly one DREAM.3D run (coordinated with MPI) that is independent of all others. Further details were reported elsewhere [TUC 16]. However, it is important to keep in mind the serial memory requirements, so that a single node is able to support the memory requirement of a single instance of DREAM.3D.

2.2.2.1. *Effect of MTR volume fraction*

Our first calculations assessed the effect of MTR volume fraction on the density of critical microstructural neighborhoods and the maximum observed hard MTR size. For this purpose, we simulated 1,500 domains measuring 100^3 voxels at 0.1875 mm/voxel resolution of quasi-unimodal, equiaxed MTRs with a mean diameter of 1.5 mm and random texture. These structures were analyzed using the method described in section 2.2.3. In addition, each MTR was assigned an intensity value meant to represent the combined role of MTR size, density and c-axis misalignment. Judiciously thresholding these data allows the number of critical neighborhoods per unit volume to be determined. The 100% MTR microstructures were also downsampled to MTR volume fractions of 0.10, 0.25 and 0.50 using random processes to convert entire MTRs to a background phase until the desired volume fraction was reached. This phase was not then considered in subsequent critical neighborhood calculations. As discussed elsewhere, an artifact with the synthetic builder during these runs necessitated removing 20 layers of the structure for a total volume of 7,908 cm^3 (~483 in^3).

2.2.2.2. *Synthetic building/extreme value theory*

Identical statistics were used to build an additional one thousand instantiations of microstructure, except this time on a 500 x 500 x 500 voxel grid yielding a total of ~$1.25x10^{11}$ voxels at 0.1875 mm resolution corresponding to an interrogated volume of ~0.823 m^3 (823 000 cm^3) or ~29 ft^3 (104x larger than the volume simulated by the 100^3 structures). The HPC microstructures were only analyzed at the fully dense (100% MTR) state. The complete dataset was taken as the ground truth against which EVT extrapolations of subsets of the data were statistically compared via bootstrapping.

2.2.3. *Crystallographic analysis for titanium dwell fatigue*

The critical microstructure for early dwell fatigue failure is characterized by the presence of neighboring (elastically and plastically) hard and soft MTRs. The magnitude of the life debit is related to the size of each constituent. Large soft MTRs lead to earlier nucleation [KIR 08], while larger hard grains lead to reduced crack growth lifetime [PIL 14]. Based on experimental observations [SIN 06, PIL 11], we can classify hard and soft regions according to their dominant c-axis orientation with the former having a $\{10\bar{1}7\}$ pole between 0 and 20° from the stress axis and the latter having a dominant c-axis orientation that lay between 70 and 90° degrees from the stress axis.

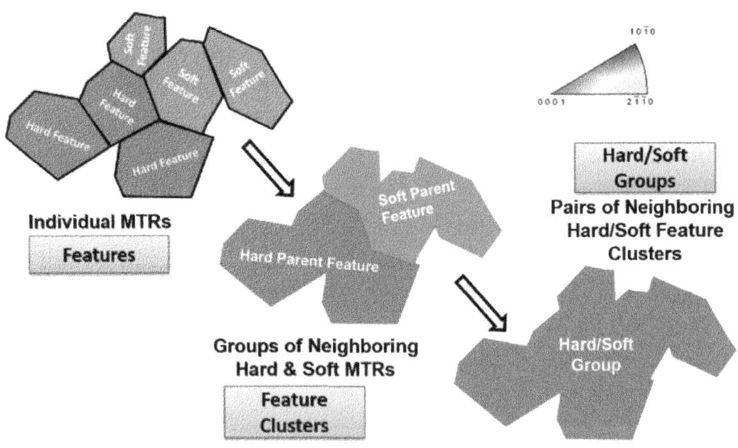

Figure 2.1. *Three levels of grouping into individual MTRs, into clusters of hard or soft MTRs and finally into hard/soft groups. For a color version of the figure, see www.iste.co.uk/brancherie/microstructure.zip*

Three levels of grouping are involved in analyzing MTRs in this study as depicted schematically in Figure 2.1. First, at the feature level, similarly oriented pixels are grouped into MTRs. Second, at the feature cluster level, groups of neighboring hard or neighboring soft MTRs are grouped with their respective classes. Finally, all hard MTRs and groups of hard MTRs are checked to see if they neighbor a soft one and these are flagged as a feature cluster pair. The weakest link is classified as the feature cluster pair with the largest total size that also maximizes hard and soft MTR size. Figure 2.2 depicts the workflow of this study. First, a stress axis is defined relative to the sample reference frame and the MTRs, in a three-dimensional (3D) microstructure instantiation, are classified as hard, soft or neither. Second, a stress axis is defined and the two-dimensional (2D) slices perpendicular to that stress axis are extracted from the 3D volume. The sizes of the feature clusters and feature cluster pairs in these 2D slices are analyzed and the size, orientation, neighbor distribution and the hard/soft MTR shared surface area are recorded. The use of 2D slices is related to experimental observations of the macroscopic fatigue cracking plane [PIL 11].

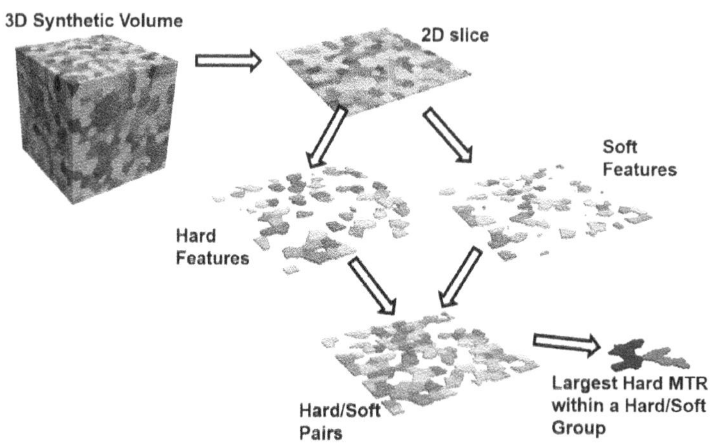

Figure 2.2. *Filtering slice-by-slice through a 3D volume. This schematic shows how the result produces the largest hard/soft pair, in 2D, in the instantiation – the weakest link. For a color version of the figure, see www.iste.co.uk/brancherie/microstructure.zip*

The full set of simulated data are analyzed to determine the density of weak links. The total dwell debit has been shown to be a strong function of the crystallographic misorientations within the MTRs

[WOO 96]. It is hypothesized that, when misorientations are, on average, high between adjacent constituents, it becomes difficult to form extended dislocation pileups, which appear to be a requisite for faceting [BRA 10] and low lifetime. Moreover, once initiated, the crack growth rates would be delayed due to the presence of high-angle boundaries. The average degree of misorientation within a MTR is represented by the intensity parameter, I_{MTR}. Experimental measurements of a similar measure of internal MTR structure [PIL 14b] suggest that this distribution decays exponentially, and thus, for this work, we have modeled it as an exponential distribution e^{-x} scaled on the interval [0,1] where 0 represents a weak MTR with a wide orientation distribution and 1 represents a crystallographically perfectly aligned MTR.

2.2.4. *Block maxima*

EVT [DON 13] provides the basis for the study of rare microstructural events. Prior applications of this branch of statistics to materials science have focused on extrapolating the distributions of individual features, like the largest inclusion size in steel [ATK 03] or the largest grain in a particular distribution [DON 13]. Since the approach described previously requires access to significant computing resources, which may not always be available, a hybrid method coupling synthetic microstructure building with EVT was employed. The synthetic builder captures the intricate details of the microstructure and the corresponding neighborhood by sampling from multiple distributions (e.g. size, orientation, c-axis misalignment and intensity) while creating the microstructure. Following the crystallographic analysis described earlier, we record only the block maxima, or the size of the single largest hard MTR that is part of a hard/soft MTR pair per microstructure. This block maxima can be extrapolated using classical EVT to determine the probability of finding a critical microstructural neighborhood in a given volume of material (with error bounds). This approach implicitly assumes that the critical microstructure can be treated as a Poisson process, that is, each observation on the input (i.e. size distribution, intensity distribution and orientation distribution) and the output (the maxima in each realization of microstructure) can be treated as independent of all others. Generalized EVT applies under these conditions and the largest hard MTR (that is also part of a hard/soft pair) expected in a given volume can be estimated by first fitting the block maxima MTR size data to the generalized extreme value distribution (GEVD, equation [2.3]) using maximum likelihood estimation (MLE) and then estimating quantiles of the model by evaluating the inverse of the GEVD with the estimated parameters (μ is the scale parameter, ξ is the shape parameter and σ is the location parameter) at a given value of interest. This value, usually called the return period, is the maximum hard MTR size (that is part of a hard/soft pair) expected to be exceeded only once in N "blocks", which occurs at the 1-1/N quantile. Here, each instantiation is a block and has a volume dictated by the voxel dimensions, allowing us to relate the largest MTR size to the total sampled volume.

$$\frac{1}{\sigma}t(x)^{\xi+1}e^{-t(x)}, t(x)=\begin{cases}\left(1+\left(\dfrac{x-\mu}{\sigma}\right)\right) if\, \xi\neq 0\\ e^{-(x-\mu)/\sigma} if\, \xi=0\end{cases}, where \begin{matrix}\mu = scale\\ \xi = shape\\ \sigma = location\end{matrix} \quad [2.3]$$

2.3. Results and discussion

The principal results of this effort include large-scale synthetic microstructure instantiations and experimentally driven segmentation and quantification of these structures for dwell-fatigue-critical microstructures. Figure 2.3 shows scatter plots that contain every hard/soft MTR pair identified in the 100^3 and 500^3 simulated volumes (7,908 cm^3 and ~0.823 m^3, respectively). These data show the hard/soft pair sizes as a function of MTR volume fraction and the size of each symbol is correlated with the shared surface area between the MTRs. The "weak links" correspond to the largest hard grains that neighbor large soft grains with substantial shared surface area. Unfortunately, the relative importance of hard and soft grain sizes on total life debit is unknown, but it is clear that larger soft MTR sizes lead to reduced initiation lifetime at constant hard MTR size while larger hard MTR sizes lead to sizable reductions in crack growth lifetime. As a quick aside, a useful study would involve calculating crack nucleation lifetime using a crystal plastic finite element model over a range of microstructural instantiations that probe the upper locus of this distribution, subsequently estimating crack growth lifetime and its associated variability [PIL 14a, PIL 16] to determine the relative importance of each microstructural feature to the total dwell debit.

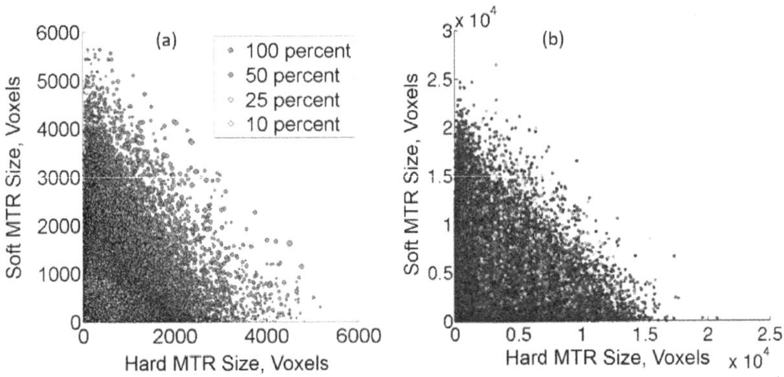

Figure 2.3. *Scatter plot of hard grain size versus soft grain size with each marker representing a hard/soft pair: (a) 100^3 and (b) 500^3 HPC structures. The marker size is proportional to the shared surface area in (a) and color in (b). The marker color in (a) corresponds to specific MTR volume percentages. For a color version of the figure, see www.iste.co.uk/brancherie/microstructure.zip*

The data generated in this study were analyzed in several ways. First, the probability of occurrence of critical microstructural neighborhoods was determined and various thresholding factors were applied to demonstrate the order-of-magnitude variation in the density of the resulting hard/soft pairs. Second, the complete distribution of hard MTR cross-sections is examined against classical distributions for various volume fractions of MTRs. Finally, the methods used for estimating the largest expected MTR in a given volume of material using EVT are described.

2.3.1. *Probability of occurrence*

In addition to visually inspecting plots like Figure 2.3 to locate potential weak links, a screening involving the joint occurrence of local microstructural features is also employed to identify weak link candidates. This exercise returns a numerical approximation of the joint probability density function (PDF) that all microstructural requirements are met.

In the present analysis, we track the MTR size distribution, orientation distribution function, MTR intensity distribution (which combines size, density, and crystallographic alignment), shared surface area distribution and, finally, an aggregate elastic modulus, *E*, for each simulated microstructure. This final parameter is a crystallite volume-weighted average of the elastic modulus of all grains in the volume as calculated from knowledge of the MTR orientation and single-crystal elastic constants for this material ($C_{11} = 136$, $C_{12} = 78$, $C_{13} = 68.5$, $C_{33} = 163$, $C_{44} = 40.6$) [KIM 02], e.g. equation [2.4]:

$$\langle E \rangle = \sum_i E_i V_i \qquad [2.4]$$

where grain *i* has orientation-dependent modulus E_i and volume V_i, and

$$E_i(\mathbf{n}) = \frac{1}{\mathbf{n}\cdot\left(\mathrm{C}_{\mathrm{ijkl}}^{-1}:(\mathbf{n}\otimes\mathbf{n})\right)\cdot\mathbf{n}} \qquad [2.5]$$

in which **n** is a unit vector in the crystal reference frame along which the modulus is calculated and C_{ijkl} is the fourth-order elastic stiffness tensor. While this is still a relatively simple approximation of the microstructure, it is easy to see how an analytical treatment of the rate of occurrence of weak links quickly becomes intractable.

By including all data points in Figure 2.3 into the rate of occurrence calculation, the approach may be considered conservative as it assumes that all pairs are "bad" even for the smallest hard MTR sizes. With additional experimental data, it will be possible to assign specific values to threshold MTRs based on size, intensity, shared surface area, local aggregate modulus or any other parameters that are determined relevant.

MTR %	Total Number (per cm^3)	Intensity > 0.9	Modulus > 123 GPa	Shared Surface Area > 1/3 MTR size
10	7	0.45	1.25	0.0008
25	23	0.60	1.05	0.04
50	34	0.65	1.55	0.40
100	30	0.62	1.25	3.20

Table 2.1. *Density of hard/soft pairs per cubic centimeter before and after application of intensity, modulus and shared surface thresholds within the 100^3 structures*

Table 2.1 illustrates that there are order-of-magnitude reductions in the number of critical MTRs for plausible values of the various criteria considered across the 100^3 structures. For total density in all volume percent MTR analyzed (i.e. 10%, 25%, 50%, and 100%), there are a striking number of critical pairs per cubic centimeter of the material – so many that we would expect far more failures to occur than are observed! Hence, it is clear that there are additional physics that must be considered, and hence, the application of mechanism-informed thresholding. The numbers trend as we would expect, with one exception in that the number density of critical MTRs at 50% is actually greater than that at 100%. This is possible because, at 100% MTR, the connectivity is so much greater that there are fewer unique MTR pairs, but the individual features comprising each feature cluster group are larger. Otherwise, the trends are intuitive. For example, for the limiting cases of 10% and 100% MTRs with random orientations, over 50,000 and less than 240,000 hard/soft MTR pairs were identified in each case, or 7 cm^{-3} and 30 cm^{-3}, respectively. Thresholding based on intensity > 0.90 decreased the number of occurrences to 3,330 and 4,905, respectively, or < 1 cm^{-3}. Excluding all MTR pairs that were contained in a box with an aggregate modulus greater than 123 GPa yielded (surprisingly) an identical number of critical pairs. The modulus threshold was in response to the work of Toubal *et al.* [TOU 09a] who reported a trend where dwell fatigue lifetime decreased with the increasing ultrasonic wave velocity, which is also proportional to elastic modulus. However, a later work by the same authors seemed to undo that correlation [TOU 09b, TOU 15]. One explanation may be as follows: in samples with high elastic modulus, there are many grains with c-axes oriented along the loading direction and, hence, this orientation has high strength. Since the fatigue tests were run at some nominal identical conditions relative to the yield strength of the tensile specimen, these samples with high modulus are actually run at a *lower* fraction of their yield strength, which is an obvious driver of longer lifetime. Furthermore, if the grains are not arranged into large MTRs but are scattered as small, discrete and hard MTRs, they may not initiate a life-limiting crack. This is why ultrasonic velocity cannot be used as a direct indicator of dwell susceptibility and more detailed analyses are required to understand the spatial clustering of grains into MTRs [PIL 14b].

MTR %	Intensity + modulus	Intensity + shared surface area	Shared surface area + modulus	Intensity + shared surface area + modulus
10	62	7908	>7098	>7098
25	36	1318	359	7908
50	33	158	52	989
100	40	14	7	304

Table 2.2. *Volume of material (in cm^3) in which only one hard/soft group that satisfies all of the criteria imposed exists (based on 100^3 structures)*

Table 2.2 explores the microstructural defect density assuming the joint occurrence of two or three different criteria. Due to the incredible reduction in the number of critical neighborhoods, these data are presented as the volume of material in cubic centimeters in which we may expect to find only one critical neighborhood. For the case of 10% MTR with three criteria imposed, there were no suitable candidates and so, at present, we know only that it requires greater than 7,908 cm^3 to locate one under the assumptions imposed.

The reported critical densities may still be conservative in that all MTR pairs were initially included in the assessment. It is likely that there may be a soft MTR size that is too small to effectively load shed, or a hard grain size that is too small to allow further crack extension if it is indeed cracked. The latter case is interesting because it links material behavior with material characterization in order to define the minimum hard MTR size that we care about. To this end, assume that the size and shape of an MTR in a plane perpendicular to the principal stress axis can be approximated as an ellipse. Now assume that this MTR is completely cracked and is subjected to a remote stress, σ. We can use the appropriate solution to solve for the stress intensity range $\Delta K = K_{max} - K_{min}$, in which $K = Y\sigma\sqrt{\pi a}$ having shape factor Y, crack length a and applied stress σ.

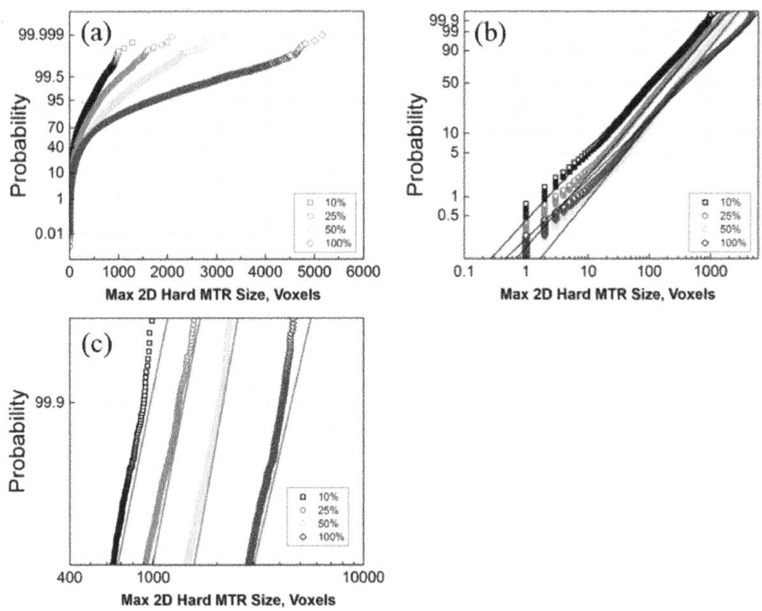

Figure 2.4. *CDF of hard MTR size plotted on (a) normal scale and (b) the Weibull scale, and also showing (c) tail departure at and immediately around the 99.9 percentile of the Weibull scale for maximum 2D hard MTR sizes for different MTR volume percentage subsets. For a color version of the figure, see www.iste.co.uk/brancherie/microstructure.zip*

2.3.2. *"Hard" MTR size distributions*

The distribution of hard grain sizes is a primary feature of interest from the perspective that, if it did not exist, there would be no early nucleation and long-range faceting leading to short lifetime. Thus, it is most critical to know the distribution of hard MTR sizes that are part of a hard/soft pair. Figure 2.4(a) shows the cumulative distribution function (CDF) of 2D hard grain size for various MTR volume percentages plotted on a normal scale. With the increasing MTR volume percentage, larger sizes are achieved and a longer upper tail is observed. Additionally, the degree of upper tail departure from normal increases with the increasing MTR percentage considered, which is more apparent in the normal scaling. As the upper tail is important with respect to predicting minimum lifetime, another model is required to capture the tail behavior. Next, the data were fit to Weibull distributions, as shown in Figure 2.4(b) with shape and scale parameters reported in Table 2.3. Here, visually, the Weibull distribution more faithfully fits the data than a normal distribution for all MTR volume percentage subsets considered (10, 25, 50 and 100% MTR), especially the upper tail, of maximum 2D hard grains. This result is interesting in light of the work of Toubal *et al.* [TOU 09b] who showed that the variability in total dwell fatigue lifetime under constant testing conditions conformed to the Weibull statistics [TOU 09b], but has a small shape parameter (beta = 2) indicating a large scatter in lifetime. This is expected in laboratory samples where the volume is relatively small and hence a limited configuration of MTRs is possible. The fact that the critical controlling microstructural unit (2D hard MTR size) follows a Weibull distribution when nothing in the synthetic building process enforces this is encouraging. While the value of Toubal *et al.* [TOU 09b] is larger than the values in Table 2.3, it is noteworthy that there is deviation of our data from Weibull's on the small tail and we have also not removed any small MTRs that may be incapable of initiating cracks (as discussed in section 2.3.1). These factors would decrease the variability in our data and hence increase the shape parameter.

MTR Volume %	Scale	Shape
10	158.9	1.08
25	244.4	1.17
50	400.7	1.25
100	523.8	0.97

Table 2.3. *Weibull scale and shape parameters reported for MTR volume percentage subsets*

2.3.3. *Block maxima*

While the Weibull distribution fits the overall 2D hard MTR size distribution well, it still did not capture the tail adequately (Figure 2.4(c)). Accurate modeling of the tail is necessary to accurately forecast rare events, and hence an alternative method of analyzing the data using the block maxima approach was pursued. In this method, the size of the single largest hard grain that is also part of a hard/soft feature cluster pair is stored per analyzed synthetic microstructure. The GEVD is fit to these maxima and the model parameters are then extrapolated to larger sample sizes. This section describes two applications of this technique.

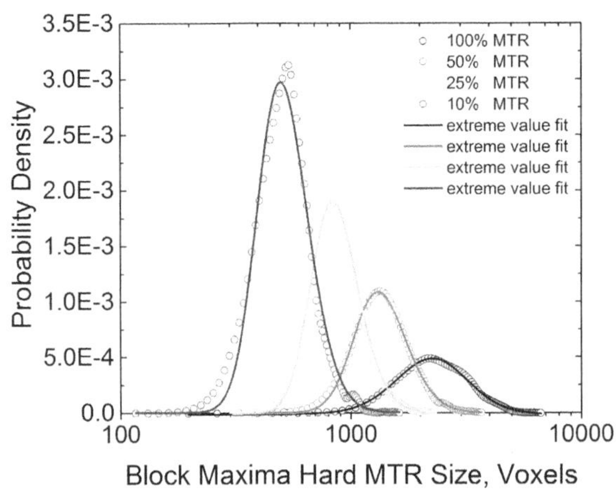

Figure 2.5. *Probability density function of block maxima of 2D hard MTR sizes with GEVD fits. For a color version of the figure, see www.iste.co.uk/brancherie/microstructure.zip*

2.3.3.1. *Extrapolation of 100^3 voxel structures*

First, we explore using the GEVD to fit the upper tail of the distribution of the 100^3 structures as a function of MTR volume fraction and then use these model parameters to estimate the largest hard MTR present in a volume that is 104x larger (equal to that simulated on the HPC). First, MLE was used to estimate the PDF of the hard MTR sizes for the 100^3 structures, the results of which are shown in Figure 2.5. The peak probability density decreases, the spread of block maxima hard grain size increases and the largest block maxima hard MTR size increases with the increasing MTR volume fraction. These three results are expected as more volume is considered and more and larger hard grains are encountered. More importantly, these fits show that the GEVD fits these distributions well giving us confidence in the model parameters obtained for extrapolation as described in section 2.2.4. The results of estimating the maximum MTR size for all MTR fractions considered to the volume of structure built on the HPC are shown in Figure 2.6(a) and compared against the 25 largest sizes obtained by the HPC simulations. Recall that only a 100% MTR structure was investigated with the HPC, and therefore, the data should only be compared against the 100% line. From these data, it is clear that the extreme value extrapolation severely underpredicts the largest critical hard MTR, and hence, we explore the confidence of the calculation in Figure 2.6(b). The confidence limits are determined by a nonlinear optimization routine, where we find a region of the GEVD parameter space that produces a log-likelihood greater than a critical value. This critical value can be estimated by evaluating the inverse chi-square distribution at the desired confidence level, e.g. 95% or otherwise. Then, to find the lower confidence bound, with the return period as the objective function, we optimize to find the smallest value under the constraint that the negative log-likelihood be less than the critical value is satisfied. The upper bound is found by simply reversing the sign of the objective function. As shown in Figure 2.6(b), the HPC data are still not captured even at the strictest level of confidence (0.9999 or 1 in 10,000), which has the widest error bounds (resulting in a factor of ~2 uncertainty in the prediction). We interpret this as being related to the different effective cross-sectional areas in the 100^3 voxel and the HPC 500^3 voxel structures. The total number of voxels in a given cross-section restricts the maximum MTR size that can be realized in that cross-section and hence it is impossible. With only two data points, there was no obvious lineal, areal or volumetric scaling between the maximum sizes determined in the two simulated volumes.

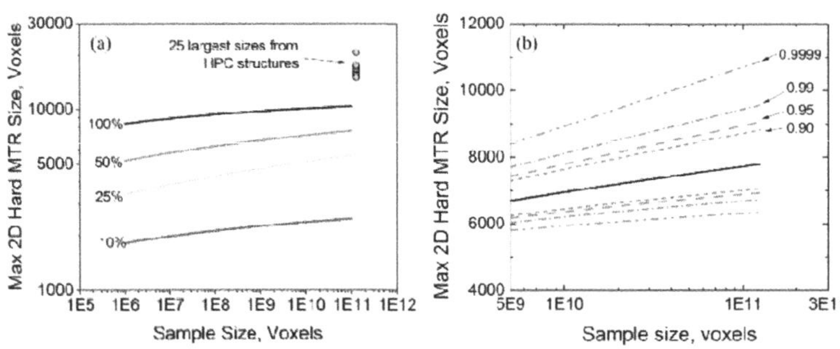

Figure 2.6. *(a) Using the GEVD parameters from the 100^3 simulated structures to predict the maximum 2D hard MTR size encountered in a volume as large as that simulated with HPC, and (b) variability in confidence limits for the case of 100% MTR. For a color version of the figure, see www.iste.co.uk/brancherie/microstructure.zip*

2.3.3.2. *Bootstrapping of 500^3 voxel structures*

In this section, to overcome the shortfall determined in the previous section, we use the HPC data as a self-consistent method of validating the approach through the application of a bootstrapping technique where we analyze subsets of the HPC data and then predict the occurrence of the largest features in the experimental distribution. To this end, we repeatedly subsampled from the available hard MTR maxima (subsample sizes were 50, 100 and 250 synthetic volumes) and converged on a set of GEVD parameters for each subset. The CDF of these distributions are shown in Figure 2.7, and the parameters obtained by fitting these data to the GEVD were used to extrapolate the upper tail and predict the largest observed MTR within the actual HPC-simulated volume. A total of one thousand 500^3 voxel structures were created and analyzed as described in section 2.3.3. 50,

100 or 250 maxima were randomly selected 50 times each to form the basis of the extreme value data which we then attempted to extrapolate (individually) to recover the value observed by the HPC. We performed similar sampling to determine the average and standard deviation of the maximum 2D hard MTR size observed for various subsamplings of the simulated structures. A few key results can be observed in Figure 2.8. The EVT predicted maximum MTR sizes and the HPC-simulated largest sizes are in good agreement (within one standard deviation) for the case where the maxima were obtained by subsampling the HPC data up to 750 instantiations, provided that a minimum of 100 block maxima were used. The change in the prediction is appreciably lower for the case when only 50 maxima are included, while increasing to 1000 maxima provides a small change relative to 250. However, increasing the number of block maxima included during MLE of the model parameters *significantly* reduces the uncertainty in the estimate. EVT extrapolation can account for all of the points in the data except for a single one. The results of the 25 largest hard MTRs observed are shown as circular data points for the case of 1,000 instantiations of 500^3 structures. All except one of these data points lie conservatively beneath the predictions based on 100 and 250 observations of maxima and the majority fall beneath the prediction for 50 maxima. A single outlier, however, with an MTR area approaching 21,000 voxels, was larger than the prediction, but did fall well within the confidence bounds considered. While it did pass the criteria imposed here, further investigation of this large outlier is warranted to see if it is a physically viable arrangement of microstructure, or if it was a fortuitous chance of connectivity across the domain, e.g. two large islands joined by a single pixel. Additionally, as evident in Figure 2.3(b), the soft MTR size for this large hard feature is perhaps too small to adequately load shed and nucleate a crack in the large hard MTR, and hence, this feature may be rejected once a minimum soft MTR size criterion is developed. Finally, it is noteworthy that this point is not captured by the GEVD extrapolation even when it is included in determining the model parameter estimates (namely the 1,000 x 500^3 line on Figure 2.8)!

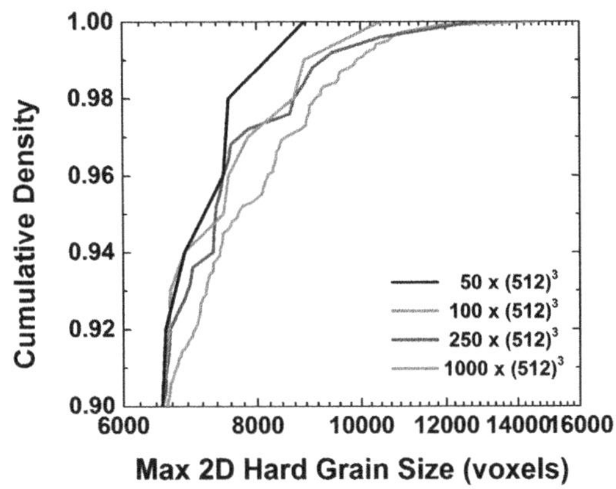

Figure 2.7. *CDF of 2D max hard MTR size obtained by subsampling block maxima from the HPC data for the volumes indicated (50, 100 and 250). For a color version of the figure, see www.iste.co.uk/brancherie/microstructure.zip*

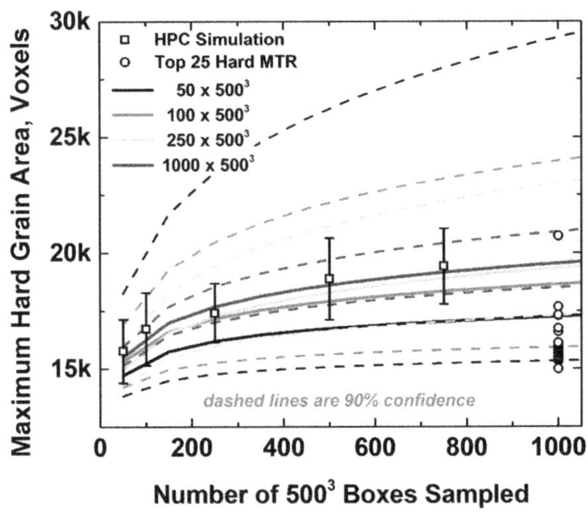

Figure 2.8. *Comparison of combined EVT extrapolation and synthetic building versus HPC simulation (see text for additional details). For a color version of the figure, see www.iste.co.uk/brancherie/microstructure.zip*

2.4. Summary and outlook

In summary, as we continue to demand more from gas turbine engines, and push materials to their useful limit, probabilistic approaches will need to be adopted which consider the role of microstructure in the fatigue and fracture process. Thus, a strong need exists to assess the probability that rarely occurring microstructural configurations will manifest within a given volume of material. In most cases, the complexity of microstructure and in particular microstructure descriptors (e.g. the ODF has more than 500 coefficients) far outweighs what can be treated analytically using previous methods. The present work describes an extensible framework for solving microstructure-based extreme value problems. Other applications of this method may include simulating as-large-as ("ALA") grains or surface-connected non-metallic inclusions, among others. The major contribution is an alternative method to numerically solving this problem using digital representations of microstructure and targeted analysis routines that are based on experimental characterization and understanding of the relevant physics. This work has made the following contributions toward this end:

– DREAM.3D was ported to the DoD DSRC platform Predator and has successfully run in an embarrassingly parallel mode to generate many statistically equivalent microstructures.

– Filters were developed in DREAM.3D to quickly interrogate the synthetic microstructures for microstructural weak links as defined by characterization of failed samples.

– A hybrid approach between synthetic building and EVT was developed, but requires additional validation.

– An estimate of the probability of occurrence of hard/soft MTR pairs was made for various volume fractions of MTRs under a variety of simplifying assumptions. Applying plausible thresholds based on other known mechanisms further reduced the number of critical microstructural feature combinations to reasonable numbers.

2.5. Bibliography

[ANA 11] ANAHID M., SAMAL M., GHOSH S., "Dwell fatigue crack nucleation model based on crystal plasticity finite element simulations of polycrystalline titanium alloys", *Journal of the Mechanics of Physics and Solids*, vol. 59, pp. 2157–2176, 2011.

[ATK 03] ATKINSON H.V., SHI G., "Characterization of inclusions in clean steels: a review including the statistics of extreme methods", *Progress in Materials Science*, vol. 48, pp. 457–520, 2003.

[BRA 10] BRANDES M.C., MILLS M.J., WILLIAMS J.C., "The influence of slip character on the creep and fatigue fracture of an alpha Ti-Al alloy", *Metallurgical and Materials Transactions A*, vol. 41, pp. 3463–3472, 2010.

[DON 13] DONEGAN S., TUCKER J., ROLLETT A. *et al.*, "Extreme value analysis of tail departure from log-normality in experimental and simulated grain size distributions", *Acta Materialia*, vol. 61, no. 15, pp. 5595–5604, 2013.

[DUN 07] DUNNE F., WALKER A., RUGG D., "A systematic study of HCP crystal orientation and morphology effects in polycrystal deformation and fatigue", *Proceedings of the Royal Society of London A*, vol. 463, pp. 1467–89, 2007.

[EVA 84] EVANS W., "Low temperature creep and fracture of near alpha titanium alloys", *Proceedings of the 2nd International Conference on Creep and Fracture of Engineering Materials and Structures*, pp. 395–406, 1984.

[EVA 94] EVANS W., BACHE M., "Dwell sensitive fatigue under biaxial loads in the near alpha titanium alloy IMI685", *International Journal of Fatigue*, vol. 16, no. 7, pp. 443–452, 1994.

[GRO 14] GROEBER M., JACKSON M., "DREAM.3D: a digital representation environment for the analysis of microstructure in 3D", *Integrating Materials and Manufacturing Innovation*, vol. 3, no. 5, 2014. doi:10. 1186/2193-9772-3-5.

[KIM 02] KIM J.-Y., YAKOVLEV V., ROKHLIN S.I., "Line-focus acoustic microscopy of Ti-6242 alpha/beta single colony: determination of elastic constants", in THOMPSON D.O., CHIMENTI D.E. (eds), *CP615, Review of Nondestructive Evaluation,* American Institute of Physics, Melville, NY, vol. 21, 2002.

[KIR 08] KIRANE K., Grain level criterion for dwell fatigue crack nucleation in polycrystalline Ti-6242 by polycrystal plasticity FE model, Masters Thesis, The Ohio State University, 2008.

[NYE 85] NYE J.F., *Physical Properties of Crystals: Their Representation by Tensors and Matrices*, Oxford University Press, New York, 1985.

[PIL 11] PILCHAK A.L., WILLIAMS J., "Observations on the role of hydrogen in facet formation", *Metallurgical and Materials Transactions A*, vol. 42, pp. 1000–1027, 2011.

[PIL 13] PILCHAK A.L., "Fatigue crack growth rates in alpha titanium: faceted vs. striation growth", *Scripta Materialia*, vol. 68, pp. 277–280, 2013.

[PIL 14a] PILCHAK A.L., "A simple model to account for the role of microtexture on fatigue and dwell fatigue lifetimes of titanium alloys", *Scripta Materialia*, vol. 74, pp. 68–71, 2014.

[PIL 14b] PILCHAK A.L., LI J., ROKHLIN S.I., "Quantitative comparison of microtexture in near-alpha titanium measured by ultrasonic scattering and electron backscatter diffraction", *Metallurgical and Materials Transactions A*, vol. 45, pp. 4679–4697, 2014.

[PIL 15] PILCHAK A.L., HUTSON A., PORTER W.J. *et al.*, "On the cyclic fatigue and dwell fatigue crack growth response of Ti-6Al-4V", in VENKATESH V., PILCHAK A.L., ALLISON J.E. *et al.* (eds), *Proceedings of the 13th World Congress on Titanium 2015*, John Wiley and Sons, 2016.

[QIU 14] QIU J., MA Y., LEI Y. *et al.*, "A Comparative Study on Dwell Fatigue of Ti-6Al-2Sn-4Zr-xMo (x= 2 to 6) Alloys on a Microstructure-Normalized Basis", *Metallurgical and Materials Transactions A*, vol. 45, no. 13, pp. 6075–6087, 2014.

[SIN 06] SINHA V., MILLS M., WILLIAMS J. *et al.*, "Observations on the faceted initiation site in the dwell-fatigue tested Ti-6242 alloy: crystallographic orientation and size effects", *Metallurgical and Materials Transactions A*, vol. 37, no. 5, pp. 1507–1518, 2006.

[SON 88] SONG Z., HOEPPNER D., "Dwell time effects on the fatigue behavior of titanium alloys", *International Journal of Fatigue*, vol. 10, no. 4, pp. 211–218, 1988.

[SON 89] SONG Z., HOEPPNER D., "Size effect on the fatigue behavior of IMI 829 titanium alloy under dwell conditions", *International Journal of Fatigue*, vol. 11, no. 2, pp. 85–90, 1989.

[TOU 09a] TOUBAL L., BOCHER P., MOREAU A., "Dwell fatigue life of a near-alpha titanium alloy and ultrasonic measurement correlation", *Proceedings of the 12th International Conference on Fracture,* Ottawa, Canada, July 12–17, 2009.

[TOU 09b] TOUBAL L., BOCHER P., MOREAU A., "Dwell-fatigue life dispersion of a near alpha titanium alloy", *International Journal of Fatigue*, vol. 31, pp. 601–605, 2009.

[TOU 15] TOUBAL L., BOCHER P., MOREAU A., "Bimodal dwell-fatigue Weibull distribution of forged titanium IMI 834", *International Journal of Fatigue*, vol. 24, no. 5, 2015. doi:10.1177/1056789514541823.

[TUC 12] TUCKER J., CHAN L., ROHRER G. *et al.*, "Tail departure of log-normal grain size distribution in 3D synthetic microstructures", *Metallurgical and Materials Transactions A*, vol. 43, no. 8, pp. 2810–2822, 2012.

[TUC 16] TUCKER J., GROEBER M., SEMIATIN S. *et al.*, "Synthetic building and targeted analysis of life-limiting microtextured region", in VENKATESH V., PILCHAK A.L., ALLISON J.E. *et al.* (eds), *Proceedings of the 13th World Congress on Titanium 2015*, John Wiley and Sons, 2016.

[VEN 06] VENKATARAMANI G., DEKA D., GHOSH S., "Crystal plasticity based FE model for understanding microstructural effects on creep and dwell fatigue in Ti-6242", *Transactions of the ASME*, vol. 356, pp. 356–365, 2006.

[WIL 02] WILLIAMS J.C., BAGGERLY R.G., PATON N.E., "Deformation behavior of HCP Ti-Al alloy single crystals", *Metallurgical and Materials Transactions A*, vol. 33, pp. 837–850, 2002.

[WOO 96] WOODFIELD A., GORMAN M., CORDERMAN R. *et al.*, "Effect of microstructure on dwell fatigue behaviour of Ti-6242", *Titanium '95: Science and Technology*, Institute of Materials, Birmingham, UK, pp. 1116–23, 1996.

3

Wave Propagation Analysis in 2D Nonlinear Periodic Structures Prone to Mechanical Instabilities

In this chapter, we analyze wave propagation in periodic network materials showing higher-order gradient effects in the presence of a nonlinear response. For this purpose, second-order gradient nonlinear continua are evaluated as substitution media for the initially discrete networks. We determine the effective mechanical energy of pantographic extensible lattices in the nonlinear regime, considering extensible situations and following the heuristic homogenization procedure already used by Piola. The network is modeled as a set of connected beams obeying a nonlinear kinematics.

On the basis of the energy expression of the identified hyperelastic second-grade continuum, we analyze the propagation of longitudinal waves in the homogenized nonlinear pantograph microstructure, which is incorporated into 1D periodic macroscopic beams as a first step, and then we study the 2D case. The obtained second-order gradient nonlinear continuum has two propagation modes (subsonic and supersonic), in contrast to Cauchy or micropolar continua, which only exhibit supersonic modes characterized by an increase of frequency with the wavenumber.

Chapter written by Hilal REDA, Yosra RAHALI, Jean-François GANGHOFFER and Hassan LAKISS.

For a weak nonlinearity, a supersonic mode occurs and the dispersion curves lie above the linear dispersion curve ($v = v_0$). For a higher nonlinearity, the wave changes from a supersonic to an evanescent subsonic mode at s = 0.8, the dispersion curves drop below the linear one and the frequency vanishes for certain values of wavenumber.

3.1. Introduction

Structures having a periodic distribution of their geometry and material properties of their constituents present interesting wave propagation properties, such as the existence of frequency band gaps, local resonances, response directionality due to their anisotropy, left-handedness, cloaking or negative acoustic refraction. The mechanical response of periodic networks has fostered much research activity in the literature, but the analysis of their dynamic and acoustic properties remains a scientific challenge [LIU 00, GOF 02]. The dynamic response and wave propagation properties of periodic networks have raised numerous studies especially in aeronautics and automotive industry, in order to reduce or absorb vibrations, impact and sound in structural components [RED 16a, RED 16b, RED 16c]. The continued interest for developing lattice materials is accompanied by recent advances in manufacturing techniques like 3D additive printing, allowing the production of lattice materials with perfect control of their microstructure and at low cost.

Studies devoted to the modeling of such lattice materials mostly restrict to the geometrically linear regime; in many applications, however, the design of bending-dominated lattices exploited for morphing structures [ASH 00, WAN 07, SPA 07] requires models of lattice materials in the nonlinear regime. We currently develop dedicated homogenization methods for network materials consisting of beam elements, advanced in [DOS 12], which shall be extended to the nonlinear regime in this study. We assume due to the small bending stiffness of the structural beam elements that the nonlinear response of the network is essentially due to the change of the network configuration (the orientation and length of the beams change with ongoing deformation). The geometrical nonlinear behavior of cellular structures and network materials was extensively studied in [WAR 89, WAR 91], considering the example of foams, using simple pin-jointed models for which the bending contribution of the skeleton struts was neglected. The authors in [WAN 14] proposed another approach where axial, bending and twisting deformations at local level were considered. More recently, [JAN 03] extended the linear model developed in [JAN 05, JAN 11] to construct the stress–strain relation and strain energy function for hyperelastic cellular materials with an arbitrary symmetry. An alternative approach was proposed by [VIG 14] using a computational homogenization method to derive a nonlinear constitutive model for network materials.

The incorporation of nonlinear aspects of wave propagation in structures is necessary whenever large deformations occur [BHA 79, OGD 97, NOR 99, POR 03], but it remains a considerable challenge. Two types of nonlinearity may be present in a broad sense: material nonlinearity [LUR 90, MAN 13] and geometrical nonlinearity [PAR 06]. The latter type of nonlinearity is related to the evolution of the microstructure or structure configuration, such as the change of configuration of a repetitive network, and it can be modeled as a succession of incremental deformations promoting continuous changes of the structure geometry. The presence of a nonlinearity in periodic structures results in amplitude-wave dependency in the dispersion relations; this opens new possibilities for a passive tuning of the dispersion band structure through an amplitude dependency of propagating waves, thereby going beyond a mere control of the dynamic and acoustic properties of repetitive structures by the design [MAN 13].

Nonlinear periodic networks support a variety of wave solutions depending on wave amplitude, wave interactions and the type of nonlinearity, for example, solitary wave solutions for Boussinesq-type equations, harmonic plane wave or discrete breathers [DAR 06, SEN 08, VAK 94, RED 16a, RED 16b].

The range of analysis of wave propagation is limited in the literature by stability considerations. Various requirements (Baker–Ericksen inequalities, Hadamard strong ellipticity condition, etc.) are often imposed to the constitutive models to study the appearance of instable behaviors. Material instability implies that no wave propagation occurs for certain directions; thus, certain modes will vanish for specific wave numbers k. The strong-ellipticity condition developed in [BAR 94] will be used here to analyze the conditions for material instabilities and their impact on wave propagation. The propagation of nonlinear elastic waves in such instable materials has been less studied in the literature, which motivates the present study.

In this chapter, we analyze the dynamic properties of extensible periodic pantographic lattices in the nonlinear regime, considering geometrical nonlinearities at the micro level. We shall focus on the pantographic network, which by essence is prone to second-gradient effects. A hyperelastic second-gradient-effective continuum is constructed for this network in section 3.2, which is at the base of subsequent wave propagation analyses. The dynamic equilibrium equation for a one-dimensional macro beam incorporating a pantographic network is written in terms of the Cauchy stress and hyperstress in a 1D context (section 3.2), which is then further recast into an ordinary differential equation by an adequate change of variables (section 3.3). The Legendre–Hadamard stability condition is expressed in section 3.3 in order to explain the observed evanescent subsonic mode. Wave propagation is next analyzed in a 2D context for the same pantographic network, highlighting similar wave characteristics, with a pronounced dispersion due to second-gradient effects and anisotropy due to nonlinearity.

3.2. Extensible energy of pantograph for dynamic analysis

In order to set the stage and in view of the dynamic analysis, we determine the energy density of the pantographic networks considered as a network of nonlinear beam elements.

3.2.1. *Expression of the pantographic network energy*

The general expression in the nonlinear framework, of the forces and moment exerted on a beam element (Figure 3.1), is derived after lengthy calculation and considering only the first-order expansion of the Taylor series of the trigonometric [BAT 79] as:

$$N = EA\left(\frac{\Delta\mathbf{U}.\mathbf{e}^b}{L}+\frac{1}{2}\left(\frac{\left(\Delta\mathbf{U}.\mathbf{e}^b\right)^2}{L^2}+\frac{\left(\Delta\mathbf{U}.\mathbf{e}^{b\perp}\right)^2}{L^2}\right)+\frac{\psi_c}{2}\right)+EI_Z\left(\frac{\Delta\psi_c}{2}\right)^2 \quad [3.1]$$

$$T = GA\left(-\psi_c\left(\frac{\Delta\mathbf{U}.\mathbf{e}^b}{L}+1\right)+\frac{\Delta\mathbf{U}.\mathbf{e}^{b\perp}}{L}\right) \quad [3.2]$$

$$M = EI_z\left(\frac{\Delta\psi_c}{L}\left(1+\frac{\Delta\mathbf{U}.\mathbf{e}^b}{L}\right)+\psi_c\frac{\Delta\mathbf{U}.\mathbf{e}^{b\perp}}{L}\right) \quad [3.3]$$

where A is the beam cross-section, L is the beam length, I_z is the quadratic moment of the considered beam, $GA=\frac{12EI_Z}{L^2}$ is the shear rigidity, ψ_c is the rotation of the central node of the beam, vectors $\mathbf{e}^b,\mathbf{e}^{b\perp}$ are, respectively, the unit director of the beam and the transverse unit vector and vector $\mathbf{U}$ is a two-dimensional displacement field parameterized by the curvilinear coordinate denoted as s= (s_1, s_2), with components $\mathbf{U}=(\xi,\eta)$.

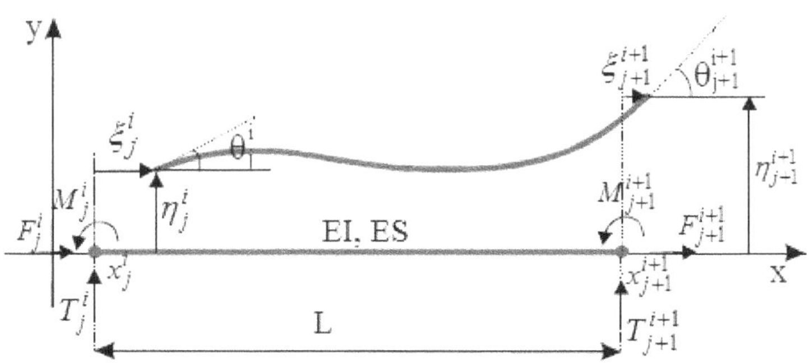

Figure 3.1. *Kinematics of a beam element*

Those expressions are written in a general case and exhibit a nonlinear elastic behavior due to the consideration of a quadratic measure of the displacement gradient along the beam and transverse to it, traducing geometrical nonlinearities in extension and flexion at beam level. In order to apply the asymptotic homogenization method, some simplifications are made.

We consider that all beams are initially rectilinear, so that no initial deformations or initial curvatures are present, implying that the corresponding kinematic variable $\Delta\psi_c$ is nil in the expression of the normal force (i.e. we neglect the coupling between tension and bending). The unit cell of the pantograph network made of two extensible beams linked at a pivot is depicted in Figure 3.2.

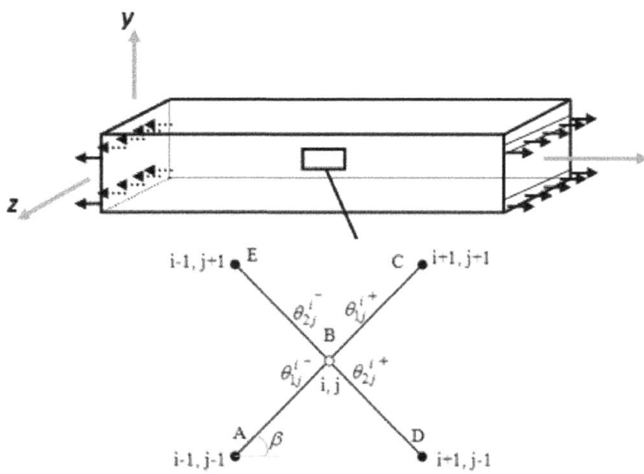

Figure 3.2. *Beams incorporating many unit cells of the pantograph network. The link at node (i,j) is a pivot*

Both beams in Figure 3.2 linking together the pair of nodes (i−1, j−1)−(i+1, j+1) and (i+1, j−1) − (i−1, j+1) have shear rigidities denoted by K_{c_1} and K_{c_2}, flexural rigidities denoted by K_{f_1} and

K_{f_2} and extensional rigidities denoted by K_{l_1} and K_{l_2}; these two beams are articulated by a pivot at the central node. We make the assumption that the two beams have equal rigidities.

On the basis of the general beam equation, the expression of the internal deformation energy can be written in intrinsic format (using the curvilinear coordinates attached to the beam) as:

$$W=\left(\begin{array}{l}\frac{1}{2}\frac{ES}{L}\left[\begin{array}{l}\left(\xi_{j+1}^{i+1}-\xi_j^i\right)+\frac{1}{2L}\left(\xi_{j+1}^{i+1}-\xi_j^i\right)^2\\+\frac{1}{2L}\left(\eta_{j+1}^{i+1}-\eta_j^i\right)^2+\frac{L}{2}\left(\theta_{j+1}^{i+1}+\theta_j^i\right)\end{array}\right]^2+\frac{1}{2}\frac{EI}{L}\left[\begin{array}{l}\left(\theta_{j+1}^{i+1}-\theta_j^i\right)\\+\frac{1}{L}\left(\begin{array}{l}\left(\theta_{j+1}^{i+1}-\theta_j^i\right)\\\left(\xi_{j+1}^{i+1}-\xi_j^i\right)\end{array}\right)\end{array}\right]^2\\+\frac{1}{2}\frac{12EI}{L^3}\left[\left(\eta_{j+1}^{i+1}-\eta_j^i\right)-\frac{L}{2}\left(\theta_{j+1}^{i+1}+\theta_j^i\right)-\left(\frac{1}{2}\left(\xi_{j+1}^{i+1}-\xi_j^i\right)\left(\theta_{j+1}^{i+1}+\theta_j^i\right)\right)\right]^2\end{array}\right) \quad [3.4]$$

This energy is made of three terms successively representing the different forms of strain energy, namely extensional, flexural and shear contributions, introducing therein the extensional rigidity, $K_l=\frac{ES}{L_b}$; the flexural rigidity, $K_f=\frac{EI}{L_b}$; the shear rigidity, $K_c=\frac{12EI}{L_b^3}$ and the beam length, $L_b=a$.

On the basis of [RAH 15] and using equation [3.4], we evaluate the energy density of the pantographic lattice (Figure 3.2, shown in the Appendix (section 3.5)). We next analyze one-dimensional and two-dimensional wave propagation successively.

3.2.2. *Dynamic equilibrium equation*

For extensible beams, previous developments show that the energy density of a nonlinear second-gradient medium takes the more general form:

$$Ws=Ws_1+Ws_2+Ws_3 \quad [3.5]$$

where Ws_1 and Ws_2 are the first- and second-order linear contributions to the energy density in small strain and Ws_3 is the contribution of the energy density accounting for the presence of large strains.

The first-order Cauchy stress is next derived in component form as follows:

$$\sigma_{ij} = \frac{\partial\left(Ws_1 + Ws_3\right)}{\partial\left(\partial u_i \big/ \partial x_j\right)} \qquad [3.6]$$

The second-order linear hyperstress follows similarly as:

$$S_{ijk} = \frac{\partial\left(Ws_2\right)}{\partial\left(\partial^2 u_{ij} \big/ \partial x_k^{\;2}\right)} \qquad [3.7]$$

The equations of motion for a second-gradient medium along the x_j direction are written in index format as follows:

$$\left(\frac{\partial \sigma_{ij}}{\partial x_j}\right) - \frac{\partial^2 S_{ijk}}{\partial x_j \partial x_k} = \rho^* \ddot{u}_j \qquad [3.8]$$

The effective density therein is given in general by $\rho^* = \frac{M_l}{A_{cell}}$, where M_l is the mass of the set of lattice beams and A_{cell} is the area of the periodic unit cell shown in Figure 3.2.

From equation [3.8], we obtain a set of differential equations that describe the propagation of longitudinal waves polarized in the direction of incident wave and of shear waves polarized in a perpendicular direction. The longitudinal and shear modes are coupled via the linear (for Cauchy and hyperstress) and nonlinear kinematic parameters.

3.3. Wave propagation in a nonlinear elastic beam

We study the propagation of longitudinal waves in a one-dimensional continuum beam at the macro level (Figure 3.2), incorporating many pantographic unit cells repeated by periodicity along the longitudinal direction, but with only one unit cell in the thickness direction, in the case of a uniaxial state of loading ($\sigma_{xx} \neq 0$ and $\varepsilon_{xx} \neq 0$).

The extensible energy density of the pantograph in the x-direction is obtained following the steps described in the Appendix as:

$$W = \frac{1}{2} K_l \varepsilon_{xx}^2 + \frac{a^2}{8} \left(K_l + K_c \right) \varepsilon_{xx,x}^2 + \frac{\sqrt{2}}{2} K_l \varepsilon_{xx}^3 \qquad [3.9]$$

introducing here the uniaxial strain $\varepsilon_{xx} = u_{,x}$. Here and in what follows, the comma denotes the partial derivative. From equation [3.9], we easily obtain the first-order Piola–Kirchhoff stress and hyperstress tensors on the basis of the general definition introduced in equations [3.6] and [3.7]. Inserting these expressions in the equation of motion [3.8] results in the homogenized nonlinear wave equation:

$$E_1 \frac{\partial^2 u}{\partial x^2} + E_2 \frac{\partial^2 u}{\partial x^2} \frac{\partial u}{\partial x} + E_3 \frac{\partial^4 u}{\partial x^4} = \rho^* \frac{\partial^2 u}{\partial t^2} \qquad [3.10]$$

where u is the longitudinal displacement, $E_1 = K_l$ is the linear effective modulus, $E_2 = 3\sqrt{2}\, K_l$ is the nonlinear effective modulus of the large deformation behavior and $E_3 = -\frac{a^2}{4}\left(K_l + K_c \right)$ is the second-order effective modulus. The PDE equation of motion [3.10] describes the quasi-static (non-dispersive) behavior of periodic pantographic lattices; it shows a coupling between first- and second-order gradient terms via the modulus E_2. Let us note that, according to equation [3.10], the second-order modulus E_3 is always negative. This means that the presence of a microstructure results in a positive dispersion, because an increase in the wavenumber leads to a corresponding increase in the phase velocity.

The problem under consideration in equation [3.10] is described by a Boussinesq-type equation: an analogous mathematical model has been essentially studied for shallow water waves to describe the internal wave in layered fluids [BOU 72]. The general character of Boussinesq and Boussinesq-type equations has been intensively studied including nonlinear effects in the dynamics of solids in [CHR 07, ENG 11]. We shall note that equation [3.10] can be integrated using elliptic functions. Single elliptic wave solutions for this equation are well known in the literature [BHA 79]. A set of elliptic functions depending on the degree of nonlinearity are presented explicitly in [AND 13]; they are described by a universal coefficient s, which does not depend on the mechanical properties of the material but takes into account the shape, period and velocity of waves within the material. It may be considered as a quantitative measure of how much the nonlinearity mode differs from the linear mode.

We shall next derive an explicit analytical stationary solution of the governing wave equation [3.10]. Using the change of variable $z = x - \frac{w}{k}t$, where w is the frequency and k is the wavenumber, and after a simple transformation, the PDE [3.10] is transformed into the ordinary differential equation for the new non-dimensional strain of the wave function, variable $N(z) = \frac{\partial u(z)}{\partial z}$:

$$\frac{\partial^2 N}{\partial^2 z} + aN + bN^2 + c = 0 \qquad [3.11]$$

where $a = \frac{k_l - \rho \frac{\omega^2}{k^2}}{-\frac{a^2}{4}(K_l + K_c)}, b = \frac{\frac{3}{2}\sqrt{2}K_l}{-\frac{a^2}{4}(K_l + K_c)}$ and c is an integration constant.

The solution of equation [3.11] can then be expressed as [AND 13, RED 16]:

$$N(z) = -\frac{A}{2} + \frac{As^2}{2\left(1 - E(s)/K(s)\right)} sn^2\left(\frac{k_0}{2} z, s\right), \qquad [3.12]$$

where s is the universal constant describing the degree of nonlinearity ($0 \leq s \leq 1$), sn(.) is the elliptic Jacobin sine, K(s) and E(s) are respectively the complete elliptic integrals of the first and second kind and k_0 is the propagation constant related to the wavenumber k as follows:

$$k = \frac{\pi}{2K(s)} k_0 \qquad [3.13]$$

3.3.1. *Legendre–Hadamard ellipticity condition and loss of stability*

The strong ellipticity condition will be used in connection to wave propagation problems; the loss of strong ellipticity implies that the modes of propagation based on dynamic equilibrium will vanish (nil wave speed). Hadamard inequality traducing the strong ellipticity condition is equivalent to the two independent inequalities:

$$\begin{aligned} &a.\frac{\partial(W_1)}{\partial F\,\partial F}(F).a > 0 \quad \forall F, a \neq 0 \\ &b.\frac{\partial(W_2)}{\partial F_{,x}\,\partial F_{,x}}(F_{,x}).b > 0 \quad \forall F_{,x}, b \neq 0 \end{aligned} \qquad [3.14]$$

where W_1, W_2 are the strain energies at first order and second order, respectively, a and b are constants and $F = 1 + u_{,x}$ is the transformation gradient.

The linear formulation of the second-gradient strain energy implies that the second part of equation [3.14] is always positive for any value of parameter b. Let us consider the expression of the strain energy in 1D including the linear, nonlinear and second gradient parts, equation [3.9]. For a high nonlinearity (s close to 1), there is a great influence of the nonlinear parts on the constitutive law, and previous inequality equation [3.14] becomes:

$$a.\frac{\partial\left(W_{s1}+W_{s3}\right)}{\partial F\,\partial F}\left(F\right).a=\left(E_1-E_2.F\right)a^2 \qquad [3.15]$$

The right-hand side of [3.15] becomes negative for $F>\frac{E_1}{E_2}$, in which case the strong-ellipticity condition is not satisfied. This entails the instability of the propagation of waves with zero propagation speed, thereby explaining the vanishing of the subsonic modes beyond certain values of the wavenumber k.

In the case of small values of s (s close to 0), the influence of the nonlinearity is weak, thus the constitutive law is dominated by the linear parts only; the strong-ellipticity condition then resumes to:

$$a.\frac{\partial\left(W_{s1}\right)}{\partial F\,\partial F}\left(F\right).a=\left(E_1\right)a^2>0 \qquad [3.16]$$

As shown in equation [3.16], the right-hand side is always positive; thus, the strong-ellipticity condition is satisfied.

3.3.2. *Supersonic and subsonic modes for 1D wave propagation*

We represent in Figure 3.3 the dispersion relation for different values of the nonlinear parameter s. In the sequel, v and v_0 are the phase velocities in the nonlinear and linear media, respectively. For a weak nonlinearity, a supersonic mode occurs and the dispersion curves lie above the linear dispersion curve ($v = v_0$). In this case, the strong-ellipticity condition is satisfied according to equation [3.16].

For a higher nonlinearity, waves change from a supersonic to an evanescent subsonic mode for (s = 0.8) and the dispersion curves drop below the linear one, vanishing for certain values of the wave number k, for which the strong-ellipticity condition is violated.

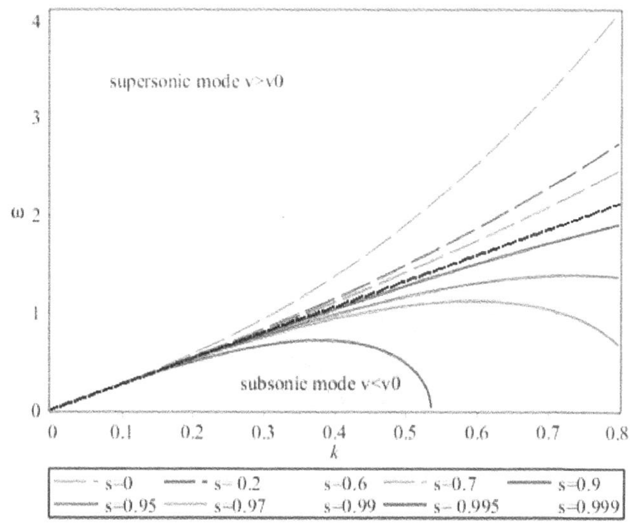

Figure 3.3. *Dispersion relation for different values of the nonlinearity parameter s. For a color version of the figure, see www.iste.co.uk/brancherie/microstructure.zip*

In the nonlinear case, the second-order gradient anisotropic continuum has two propagation modes, namely subsonic and supersonic, whereas Cauchy or micropolar continua only have supersonic modes, for which an increase of the frequency with the wavenumber occurs.

Figure 3.4 shows that the group velocity tends to zero when the wave frequency of all evanescent subsonic modes approaches their maximum value. The vanishing of mode occurs when the group velocity $v_g \to -\infty$, which means that energy is completely dissipated.

For the subsonic mode, an increase of the wavenumber k entails an accumulation of energy (increase in the group velocity), which in turn shifts the dispersion diagram toward higher frequencies.

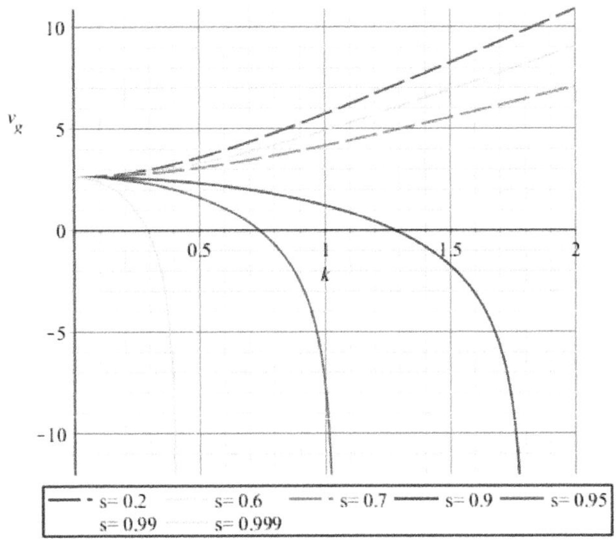

Figure 3.4. *Group velocities for different values of s. For a color version of the figure, see www.iste.co.uk/brancherie/microstructure.zip*

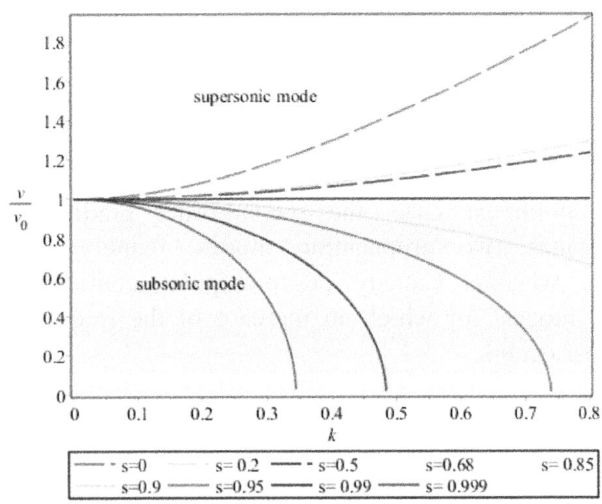

Figure 3.5. *Variation of the phase velocity for different values of s. For a color version of the figure, see www.iste.co.uk/brancherie/microstructure.zip*

Plot of the phase velocity versus wavenumber for different values of s shown in Figure 3.4 shows that for a weak nonlinearity, the phase velocity v exceeds the linear (non-dispersive) velocity v_0, describing a supersonic mode, whereas for a higher nonlinearity, the nature of the wave changes from a supersonic to a subsonic mode (approximately for s = 0.8).

We next extend the wave propagation analysis to a 2D situation.

3.3.3. *Wave dispersion relation in 2D nonlinear periodic structures*

The extensible energy density of the pantograph in a 2D plane can be written (developments are shown in the Appendix) using (x, y) Cartesian coordinates as:

$$W = \left(\begin{array}{l} \frac{1}{2} K_l \left(\left(\varepsilon_{xx} + \varepsilon_{xy} \right)^2 + \left(\varepsilon_{yy} + \varepsilon_{yx} \right)^2 \right) + \frac{a^2}{8} (K_l + K_c) \\ \left(\left(\varepsilon_{xx,x} + \varepsilon_{xy,y} \right)^2 + \left(\varepsilon_{yy,y} + \varepsilon_{yx,x} \right)^2 \right) \\ + \frac{\sqrt{2}}{2} K_l \left(\begin{array}{l} \left(2\varepsilon_{yx}\varepsilon_{xy} + \left(\varepsilon_{xx}{}^2 + \varepsilon_{yy}{}^2 \right) \right) \left(\varepsilon_{xx} + \varepsilon_{yy} \right) + \\ \left(\varepsilon_{xx}{}^3 + \varepsilon_{yy}{}^3 \right) + 3 \left(\varepsilon_{xx}\varepsilon_{xy}{}^2 + \varepsilon_{yx}{}^2 \varepsilon_{yy} \right) \end{array} \right) \end{array} \right) \quad [3.17]$$

The nonlinear wave equation in 2D can then be formulated on the basis of the two dynamic equilibrium equations involving the constitutive law expressed from the strain energy density in equation [3.17]. The methodology for determining the acoustic characteristic (the dispersion relation, the phase and group velocities) for periodic uniform 2D networks is the same as the previous one exposed in the 1D situation. We further extend equation [3.12] in 2D by taking the same form of the non-dimensional strain in y direction as in 1D, but now with two different amplitudes:

$$N(z) = -\frac{A_0}{2} + \frac{A_0 s^2}{2\left(1 - \frac{E(s)}{K(s)}\right)} sn^2\left(\frac{k_0}{2} z, s\right) = \frac{\partial u}{\partial z}$$
$$M(z) = -\frac{B_0}{2} + \frac{B_0 s^2}{2\left(1 - \frac{E(s)}{K(s)}\right)} sn^2\left(\frac{k_0}{2} z, s\right) = \frac{\partial v}{\partial z} \qquad [3.18]$$

where $z = x + y - v_p t$. Inserting the displacements into the equation of motion and canceling the determinant of the system with the two amplitude variables, we obtain a dispersion relation (relation between frequency and wavenumber) for different values of parameter s.

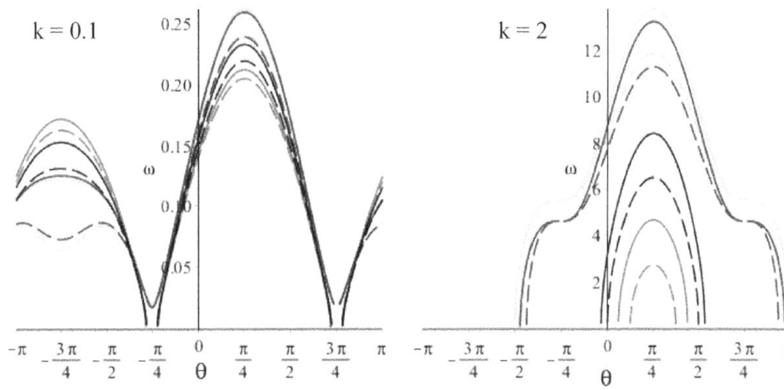

Figure 3.6. *Frequency versus direction of propagation with different values of universal constant s and for two values of the wave number (k=0.1 and 2). Color code: red for s=0.999, black for s=0.95, blue for s=0.4 and green for s=0.1. For a color version of the figure, see www.iste.co.uk/brancherie/microstructure.zip*

In Figure 3.6, the solid lines correspond to the longitudinal mode and the dashed lines to the shear mode. For lower values of the wavenumber, the two modes (supersonic and subsonic) travel in all directions of propagation. By increasing the wave number, the subsonic modes are shown to vanish (for large values of the wave number); it is expected that this range of discontinuity will increase with the wavenumber.

In the case of a higher nonlinearity, the frequency of subsonic modes increases with the wavenumber. The difference between the longitudinal and shear modes in both cases results from the large deformation aspect of the energy. By varying parameter s from a weak to a high nonlinearity, the stress and hyperstress coefficients become larger for the longitudinal mode in comparison to the transverse mode ($\sigma_{11} > \sigma_{22}$, $S_{111} > S_{222}$), resulting in a lower dispersion relation for the shear mode.

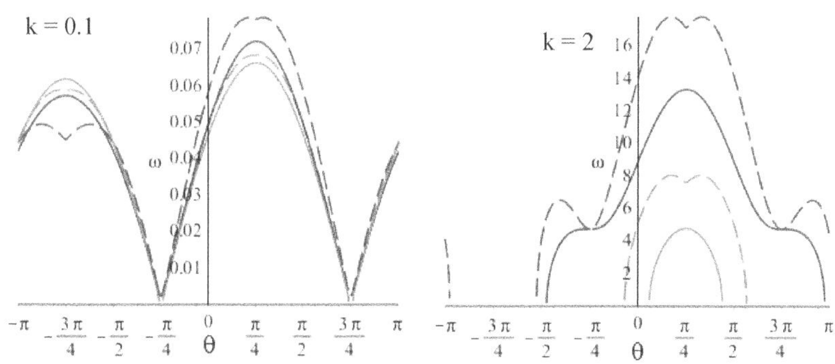

Figure 3.7. *Frequency for the longitudinal mode versus the direction of propagation for different values of the universal constant s and wavenumber (k = 0.1 and 2). Blue line: s = 0.1 (supersonic mode); red line: s = 0.99 (subsonic mode). continuous line: A_0 = B_0 (amplitude of wave) = 1; dashed line: A_0 = B_0 = 5. For a color version of the figure, see www.iste.co.uk/ brancherie/microstructure.zip*

Figure 3.7 shows the dispersion curves for different values of s versus the direction of propagation and the wave amplitude; it demonstrates that the dispersion curve increases with wave amplitude for both subsonic and supersonic modes, a phenomenon known as amplitude-dependent dispersion relation. Because of the symmetry of the pantograph structure, the same evolution of the frequency can be observed when changing the amplitude A_0 or B_0. Figure 3.7 highlights that the discontinuity of the frequency will change with the amplitude for higher values of wavenumber.

3.3.4. *Anisotropic behavior of 2D pantographic networks versus the degree of nonlinearity*

We show in Figures 3.8 and 3.9 the phase velocity versus the direction of propagation in polar form in both linear and nonlinear regimes. The main objective is to study the influence of the nonlinearity on the dispersive, anisotropy and symmetry properties of the pantograph network.

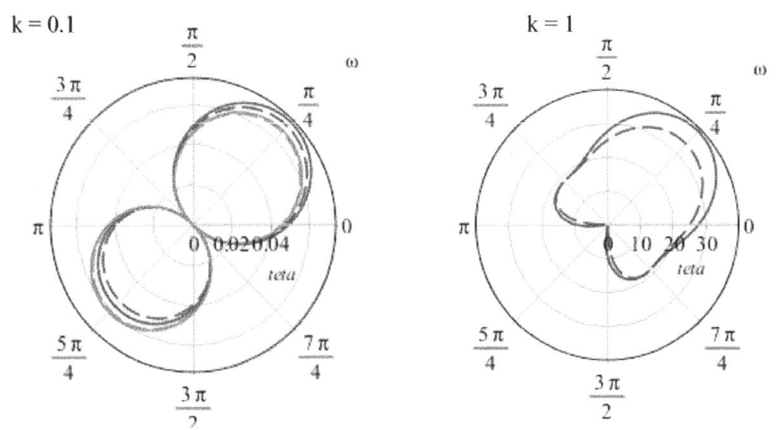

Figure 3.8. *Phase velocity of the pantograph network for the nonlinear effective medium for two values of k. Continuous line: longitudinal mode; dashed line: shear mode; red line: s = 0.95 (supersonic mode); blue line: s = 0.1 (subsonic mode). For a color version of the figure, see www.iste.co.uk/brancherie/microstructure.zip*

In the linear case (Figure 3.9), the obtained circular shape of the phase velocity implies that the pantograph network behaves in an isotropic manner; this is in contrast to the nonlinear case for which the irregular shape of the phase velocity plots implies the anisotropic behavior of wave propagation. For both linear and nonlinear situations, an increase of the wavenumber entails an increase of the modulus and shape of the phase velocity, thus reflecting the dispersive behavior for the effective second-gradient medium. We shall note here the non-dispersive nature of the effective medium in the linear case and in the absence of second-gradient terms. Another interesting aspect when moving from the linear to the nonlinear situation is the disappearance of the symmetry properties of the pantograph, especially for higher values of wavenumber. It is also worth noting that the subsonic mode disappears when the wavenumber is increased.

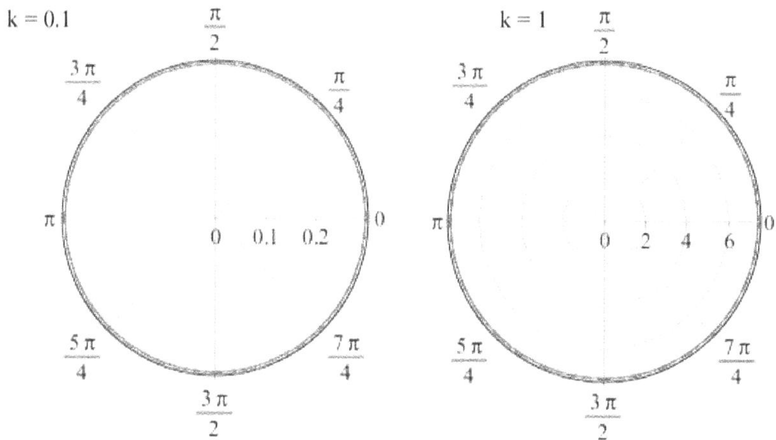

Figure 3.9. *Phase velocity of the pantograph network for the linear effective medium for two values of k. Red line: longitudinal mode; green line: shear mode. For a color version of the figure, see www.iste.co.uk/brancherie/microstructure.zip*

3.4. Conclusion

Wave propagation within nonlinear second-gradient continua has so far been devoted little attention in the literature. We have analyzed nonlinear wave propagation into periodic network materials showing second-order gradient effects, on the basis of effective hyperelastic second-gradient models. The methodology presented here is quite general and relies on dedicated homogenization techniques for such network materials. We focus here on extensible pantographic networks, which exhibit these second-order effects, in addition to first-order elasticity. The effective mechanical energy of pantographic extensible networks has been evaluated in closed form in the nonlinear regime, following the heuristic homogenization procedure already used by Piola.

On the basis of the energy expression of the identified hyperelastic second-grade continuum, we analyze the propagation of longitudinal waves in the homogenized nonlinear pantograph microstructure, which is incorporated into macroscopic 1D beams. The obtained second-order gradient nonlinear continuum has two propagation modes (subsonic and supersonic), in contrast to Cauchy or micropolar continua, which only exhibit supersonic modes characterized by an increase in the frequency with the wavenumber. For a weak nonlinearity, a supersonic mode occurs and the dispersion curves lie above the linear dispersion curve ($v = v_0$). For higher nonlinearity, the wave changes from a supersonic to an evanescent subsonic mode at s = 0.8 and the dispersion curves drop below the linear case and vanish at certain values of the wavenumber. The occurrence of such cut-off frequencies at specific values of the wave number is caused by a loss of the Hadamard strong ellipticity, itself due the applied large pre-deformation. The effective medium is dispersive because of the presence of second-order gradient terms. Wave propagation is isotropic in the linear regime, and it becomes anisotropic for nonlinear waves.

The present methodology based on the homogenization of the discrete microstructure network materials to formulate enriched effective continua is quite general and shall be extended in future contributions to different network materials prone to such second-order gradient effects.

3.5. Appendix

$$W = \left(\begin{array}{l} \frac{1}{2}\frac{ES}{L}\left[\begin{array}{l} \left(\xi_{j+1}^{i+1} - \xi_j^i\right) + \frac{1}{2L}\left(\xi_{j+1}^{i+1} - \xi_j^i\right)^2 \\ + \frac{1}{2L}\left(\eta_{j+1}^{i+1} - \eta_j^i\right)^2 + \frac{L}{2}\left(\theta_{j+1}^{i+1} + \theta_j^i\right) \end{array} \right]^2 \\ + \frac{1}{2}\frac{EI}{L}\left[\begin{array}{l} \left(\theta_{j+1}^{i+1} - \theta_j^i\right) \\ + \frac{1}{L}\left(\left(\theta_{j+1}^{i+1} - \theta_j^i\right)\left(\xi_{j+1}^{i+1} - \xi_j^i\right)\right) \end{array} \right]^2 \\ + \frac{1}{2}\frac{12EI}{L^3}\left[\begin{array}{l} \left(\eta_{j+1}^{i+1} - \eta_j^i\right) - \frac{L}{2}\left(\theta_{j+1}^{i+1} + \theta_j^i\right) - \\ \left(\frac{1}{2}\left(\xi_{j+1}^{i+1} - \xi_j^i\right)\left(\theta_{j+1}^{i+1} + \theta_j^i\right)\right) \end{array} \right]^2 \end{array} \right) \qquad [A.1]$$

The strain energy of the structure is written as follows:

$$
\begin{aligned}
W = \Bigg[& \frac{1}{2}K_l \left(\left[\begin{array}{l} \left(\xi_{1j+1}^{i+1} - \xi_{1j}^{i}\right) + \frac{1}{2a}\left(\xi_{1j+1}^{i+1} - \xi_{1j}^{i}\right)^2 \\ + \frac{1}{2a}\left(\eta_{1j+1}^{i+1} - \eta_{1j}^{i}\right)^2 + \frac{a}{2}\left(\theta_{1j+1}^{i+1} + \theta_{1j}^{i\,+}\right) \end{array} \right]^2 + \left[\begin{array}{l} \left(\xi_{1j}^{i} - \xi_{1j-1}^{i-1}\right) + \frac{1}{2a}\left(\xi_{1j}^{i} - \xi_{1j-1}^{i-1}\right)^2 \\ + \frac{1}{2a}\left(\eta_{1j}^{i} - \eta_{1j-1}^{i-1}\right)^2 + \frac{a}{2}\left(\theta_{1j}^{i\,-} + \theta_{1j-1}^{i-1}\right) \end{array} \right]^2 \right) \\
& + \frac{1}{2}K_l \left(\left[\begin{array}{l} \left(\xi_{2j-1}^{i+1} - \xi_{2j}^{i}\right) + \frac{1}{2a}\left(\xi_{2j-1}^{i+1} - \xi_{2j}^{i}\right)^2 \\ + \frac{1}{2a}\left(\eta_{2j-1}^{i+1} - \eta_{2j}^{i}\right)^2 + \frac{a}{2}\left(\theta_{2j-1}^{i+1} + \theta_{2j}^{i\,+}\right) \end{array} \right]^2 + \left[\begin{array}{l} \left(\xi_{2j}^{i} - \xi_{2j+1}^{i-1}\right) + \frac{1}{2a}\left(\xi_{2j}^{i} - \xi_{2j+1}^{i-1}\right)^2 \\ + \frac{1}{2a}\left(\eta_{2j}^{i} - \eta_{2j+1}^{i-1}\right)^2 + \frac{a}{2}\left(\theta_{2j}^{i\,-} + \theta_{2j+1}^{i-1}\right) \end{array} \right]^2 \right) \\
& \frac{1}{2}K_{f_1} \left[\begin{array}{l} \left(\left(\theta_{1j+1}^{i+1} - \theta_{1j}^{i\,+}\right)\left(1 + \frac{1}{a}\left(\xi_{1j+1}^{i+1} - \xi_{1j}^{i}\right)\right)\right)^2 \\ + \left(\left(\theta_{1j}^{i\,-} - \theta_{1j-1}^{i-1}\right)\left(1 + \frac{1}{a}\left(\xi_{1j}^{i} - \xi_{1j-1}^{i-1}\right)\right)\right)^2 \end{array} \right] \\
& + \frac{1}{2}K_{f_2} \left[\begin{array}{l} \left(\left(\theta_{2j-1}^{i+1} - \theta_{2j}^{i\,+}\right)\left(1 + \frac{1}{a}\left(\xi_{2j-1}^{i+1} - \xi_{2j}^{i}\right)\right)\right)^2 \\ + \left(\left(\theta_{2j}^{i\,-} - \theta_{2j+1}^{i-1}\right)\left(1 + \frac{1}{a}\left(\xi_{2j}^{i} - \xi_{2j+1}^{i-1}\right)\right)\right)^2 \end{array} \right] \\
& + \frac{1}{2}K_{c_1} \left[\begin{array}{l} \left(\left(\eta_{1j+1}^{i+1} - \eta_{1j}^{i}\right) - \frac{a}{2}\left(\theta_{1j+1}^{i+1} + \theta_{1j}^{i\,+}\right) + \frac{1}{2}\left(\xi_{1j+1}^{i+1} - \xi_{1j}^{i}\right)\left(\theta_{1j+1}^{i+1} + \theta_{1j}^{i\,+}\right)\right)^2 \\ + \left(\left(\eta_{1j}^{i} - \eta_{1j-1}^{i-1}\right) - \frac{a}{2}\left(\theta_{1j}^{i\,-} + \theta_{1j-1}^{i-1}\right) + \frac{1}{2}\left(\xi_{1j}^{i} - \xi_{1j-1}^{i-1}\right)\left(\theta_{1j}^{i\,-} + \theta_{1j-1}^{i-1}\right)\right)^2 \end{array} \right] \\
& + \frac{1}{2}K_{c_2} \left[\begin{array}{l} \left(\left(\eta_{2j-1}^{i+1} - \eta_{2j}^{i}\right) - \frac{a}{2}\left(\theta_{2j-1}^{i+1} + \theta_{2j}^{i\,+}\right) + \frac{1}{2}\left(\xi_{2j-1}^{i+1} - \xi_{2j}^{i}\right)\left(\theta_{2j-1}^{i+1} + \theta_{2j}^{i\,+}\right)\right)^2 \\ + \left(\left(\eta_{2j}^{i} - \eta_{2j+1}^{i-1}\right) - \frac{a}{2}\left(\theta_{2j}^{i\,-} + \theta_{2j+1}^{i-1}\right) + \frac{1}{2}\left(\xi_{2j}^{i} - \xi_{2j+1}^{i-1}\right)^2\left(\theta_{2j}^{i\,-} + \theta_{2j+1}^{i-1}\right)\right)^2 \end{array} \right] \Bigg]
\end{aligned}
\qquad [A.2]
$$

Here and in the following, indices 1 and 2 refer, respectively, to the first and second beams of the pantograph unit cell.

We denote separately both contributions to the strain energy associated to each beam:

$$
\begin{aligned}
W_1 = \Bigg[& \frac{1}{2}K_l \left(\left[\left(\xi_{1j+1}^{i+1}-\xi_{1j}^{i}\right)+\frac{1}{2a}\left(\xi_{1j+1}^{i+1}-\xi_{1j}^{i}\right)^2 + \frac{1}{2a}\left(\eta_{1j+1}^{i+1}-\eta_{1j}^{i}\right)^2+\frac{a}{2}\left(\theta_{1j+1}^{i+1}+\theta_{1j}^{i\,+}\right)\right]^2 \right. \\
& \left. + \left[\left(\xi_{1j}^{i}-\xi_{1j-1}^{i-1}\right)+\frac{1}{2a}\left(\xi_{1j}^{i}-\xi_{1j-1}^{i-1}\right)^2 + \frac{1}{2a}\left(\eta_{1j}^{i}-\eta_{1j-1}^{i-1}\right)^2+\frac{a}{2}\left(\theta_{1j}^{i\,-}+\theta_{1j-1}^{i-1}\right)\right]^2 \right) \\
& \frac{1}{2}K_f \left[\left(\left(\theta_{1j+1}^{i+1}-\theta_{1j}^{i\,+}\right)\left(1+\frac{1}{a}\left(\xi_{1j+1}^{i+1}-\xi_{1j}^{i}\right)\right)\right)^2 + \left(\left(\theta_{1j}^{i\,-}-\theta_{1j-1}^{i-1}\right)\left(1+\frac{1}{a}\left(\xi_{1j}^{i}-\xi_{1j-1}^{i-1}\right)\right)\right)^2 \right] \\
& +\frac{1}{2}K_c \left[\left(\left(\eta_{1j+1}^{i+1}-\eta_{1j}^{i}\right)-\frac{a}{2}\left(\theta_{1j+1}^{i+1}+\theta_{1j}^{i\,+}\right)-\frac{1}{2}\left(\xi_{1j+1}^{i+1}-\xi_{1j}^{i}\right)\left(\theta_{1j+1}^{i+1}+\theta_{1j}^{i\,+}\right)\right)^2 \right. \\
& \left. + \left(\left(\eta_{1j}^{i}-\eta_{1j-1}^{i-1}\right)-\frac{a}{2}\left(\theta_{1j}^{i\,-}+\theta_{1j-1}^{i-1}\right)-\frac{1}{2}\left(\xi_{1j}^{i}-\xi_{1j-1}^{i-1}\right)\left(\theta_{1j}^{i\,-}+\theta_{1j-1}^{i-1}\right)\right)^2 \right] \Bigg]
\end{aligned}
\tag{A.3}
$$

For the second beam, it holds similarly:

$$W_2 = \left[\begin{array}{l} \frac{1}{2}K_l \left(\left[\begin{array}{l} \left(\xi_{2j-1}^{i+1} - \xi_{2j}^{i}\right) + \frac{1}{2a}\left(\xi_{2j-1}^{i+1} - \xi_{2j}^{i}\right)^2 \\ + \frac{1}{2a}\left(\eta_{2j-1}^{i+1} - \eta_{2j}^{i}\right)^2 + \frac{a}{2}\left(\theta_{2j-1}^{i+1} + \theta_{2j}^{i\,+}\right) \end{array} \right]^2 + \left[\begin{array}{l} \left(\xi_{2j}^{i} - \xi_{2j+1}^{i-1}\right) + \frac{1}{2a}\left(\xi_{2j}^{i} - \xi_{2j+1}^{i-1}\right)^2 \\ + \frac{1}{2a}\left(\eta_{2j}^{i} - \eta_{2j+1}^{i-1}\right)^2 + \frac{a}{2}\left(\theta_{2j}^{i\,-} + \theta_{2j+1}^{i-1}\right) \end{array} \right]^2 \right) \\ \frac{1}{2}K_f \left[\begin{array}{l} \left(\left(\theta_{2j-1}^{i+1} - \theta_{2j}^{i\,+}\right)\left(1 + \frac{1}{a}\left(\xi_{2j-1}^{i+1} - \xi_{2j}^{i}\right)\right)\right)^2 \\ + \left(\left(\theta_{2j}^{i\,-} - \theta_{2j+1}^{i-1}\right)\left(1 + \frac{1}{a}\left(\xi_{2j}^{i} - \xi_{2j+1}^{i-1}\right)\right)\right)^2 \end{array} \right] \\ + \frac{1}{2}K_c \left[\begin{array}{l} \left(\left(\eta_{2j-1}^{i+1} - \eta_{2j}^{i}\right) - \frac{a}{2}\left(\theta_{2j-1}^{i+1} + \theta_{2j}^{i\,+}\right) - \frac{1}{2}\left(\xi_{2j-1}^{i+1} - \xi_{2j}^{i}\right)\left(\theta_{2j-1}^{i+1} + \theta_{2j}^{i\,+}\right)\right)^2 \\ + \left(\left(\eta_{2j}^{i} - \eta_{2j+1}^{i-1}\right) - \frac{a}{2}\left(\theta_{2j}^{i\,-} + \theta_{2j+1}^{i-1}\right) - \frac{1}{2}\left(\xi_{2j}^{i} - \xi_{2j+1}^{i-1}\right)^2\left(\theta_{2j}^{i\,-} + \theta_{2j+1}^{i-1}\right)\right)^2 \end{array} \right] \end{array}\right] \quad [A.4]$$

The two beams have equal rigidities.

Using different finite difference schemes and a Bernoulli kinematic assumption (for more details, see [RAH 15]), the expressions W_1 and W_2 simplify to:

$$
W_1 = \left[
\begin{aligned}
&\frac{d^2}{2}K_l \left(
\begin{aligned}
&\frac{d^4}{32}\left(\left(\left.\frac{\partial^2 \xi_1}{\partial s^2}\right|_j^i\right)^4 + \left(\left.\frac{\partial^2 \eta_1}{\partial s^2}\right|_j^i\right)^4\right) + \frac{1}{2}\left(\left(\left.\frac{\partial \xi_1}{\partial s}\right|_j^i\right)^4 + \left(\left.\frac{\partial \eta_1}{\partial s}\right|_j^i\right)^4\right) + 2\left(\left(\left.\frac{\partial \xi_1}{\partial s}\right|_j^i\right)^3 + \left(\left.\frac{\partial \eta_1}{\partial s}\right|_j^i\right)^3 + \left(\left.\frac{\partial \xi_1}{\partial s}\right|_j^i\right)^2 + \left(\left.\frac{\partial \eta_1}{\partial s}\right|_j^i\right)^2\right) \\
&+\frac{d^2}{2}\left(\left.\frac{\partial^2 \xi_1}{\partial s^2}\right|_j^i\right)^2\left(1+3\left.\frac{\partial \xi_1}{\partial s}\right|_j^i + \frac{3}{2}\left(\left.\frac{\partial \xi_1}{\partial s}\right|_j^i\right)^2\right) + \left(\left.\frac{\partial \eta_1}{\partial s}\right|_j^i\right)^2\left(\frac{d^2}{4}\left(\left.\frac{\partial^2 \xi_1}{\partial s^2}\right|_j^i\right)^2 + \left(\left.\frac{\partial \xi_1}{\partial s}\right|_j^i\right)^2 + 2\left.\frac{\partial \xi_1}{\partial s}\right|_j^i\right) \\
&+d^2\left(\left.\frac{\partial^2 \xi_1}{\partial s^2}\right|_j^i\right)\left(\left.\frac{\partial \eta_1}{\partial s}\right|_j^i\right)\left(\left.\frac{\partial^2 \eta_1}{\partial s^2}\right|_j^i\right)\left(1+\left.\frac{\partial \xi_1}{\partial s}\right|_j^i\right) \\
&+\frac{d^2}{2}\left(\left.\frac{\partial^2 \eta_1}{\partial s^2}\right|_j^i\right)^2\left(\frac{3}{2}\left(\left.\frac{\partial \eta_1}{\partial s}\right|_j^i\right)^2 + \frac{1}{2}\left(\left.\frac{\partial \xi_1}{\partial s}\right|_j^i\right)^2 + \frac{d^2}{8}\left(\left.\frac{\partial^2 \xi_1}{\partial s^2}\right|_j^i\right)^2 + \left(\left.\frac{\partial \xi_1}{\partial s}\right|_j^i\right) + \left(\left.\frac{\partial \eta_1}{\partial s}\right|_j^i\right)\right) \\
&+\left(\left.\frac{\partial \eta_1}{\partial s}\right|_j^i\right)\left(4\left.\frac{\partial \xi_1}{\partial s}\right|_j^i + 2\left(\left.\frac{\partial \xi_1}{\partial s}\right|_j^i\right)^2 + \frac{d^2}{2}\left(\left.\frac{\partial^2 \xi_1}{\partial s^2}\right|_j^i\right)^2\right)
\end{aligned}
\right) \\
&+\frac{d^2}{2}K_c \left[
\begin{aligned}
&\left(\left.\frac{\partial \eta_1}{\partial s}\right|_j^i\right)^2\left(2\left(\left.\frac{\partial \xi_1}{\partial s}\right|_j^i\right)^2 + \frac{d^2}{2}\left(\left.\frac{\partial^2 \xi_1}{\partial s^2}\right|_j^i\right)^2\right) + \frac{d^2}{2}\left(\left.\frac{\partial^2 \eta_1}{\partial s^2}\right|_j^i\right)^2 \\
&-d^2\left(\left.\frac{\partial^2 \eta_1}{\partial s^2}\right|_j^i\right)\left(\left.\frac{\partial \eta_1}{\partial s}\right|_j^i\right)\left(\left.\frac{\partial^2 \xi_1}{\partial s^2}\right|_j^i\right)
\end{aligned}
\right]
\end{aligned}
\right] \qquad [A.5]
$$

and:

$$
W_2 = \left[
\begin{aligned}
&\frac{a^2}{2}K_l \left(
\begin{aligned}
&\frac{a^4}{32}\left(\left(\left.\frac{\partial^2\xi_2}{\partial s^2}\right|_j^i\right)^4+\left(\left.\frac{\partial^2\eta_2}{\partial s^2}\right|_j^i\right)^4\right)+\frac{1}{2}\left(\left(\left.\frac{\partial\xi_2}{\partial s}\right|_j^i\right)^4+\left(\left.\frac{\partial\eta_2}{\partial s}\right|_j^i\right)^4\right)\\
&+2\left(\left(\left.\frac{\partial\xi_2}{\partial s}\right|_j^i\right)^3+\left(\left.\frac{\partial\eta_2}{\partial s}\right|_j^i\right)^3+\left(\left.\frac{\partial\xi_2}{\partial s}\right|_j^i\right)^2+\left(\left.\frac{\partial\eta_2}{\partial s}\right|_j^i\right)^2\right)\\
&+\frac{a^2}{2}\left(\left.\frac{\partial^2\xi_2}{\partial s^2}\right|_j^i\right)^2\left(1+3\left.\frac{\partial\xi_2}{\partial s}\right|_j^i+\frac{3}{2}\left(\left.\frac{\partial\xi_2}{\partial s}\right|_j^i\right)^2\right)\\
&+\left(\left.\frac{\partial\eta_2}{\partial s}\right|_j^i\right)^2\left(\frac{a^2}{4}\left(\left.\frac{\partial^2\xi_2}{\partial s^2}\right|_j^i\right)^2+\left(\left.\frac{\partial\xi_2}{\partial s}\right|_j^i\right)^2+2\left.\frac{\partial\xi_2}{\partial s}\right|_j^i\right)\\
&+a^2\left(\left.\frac{\partial^2\xi_2}{\partial s^2}\right|_j^i\right)\left(\left.\frac{\partial\eta_2}{\partial s}\right|_j^i\right)\left(\left.\frac{\partial^2\eta_2}{\partial s^2}\right|_j^i\right)\left(1+\left.\frac{\partial\xi_2}{\partial s}\right|_j^i\right)\\
&+\frac{a^2}{2}\left(\left.\frac{\partial^2\eta_2}{\partial s^2}\right|_j^i\right)^2\left(
\begin{aligned}
&\frac{3}{2}\left(\left.\frac{\partial\eta_2}{\partial s}\right|_j^i\right)^2+\frac{1}{2}\left(\left.\frac{\partial\xi_2}{\partial s}\right|_j^i\right)^2\\
&+\frac{a^2}{8}\left(\left.\frac{\partial^2\xi_2}{\partial s^2}\right|_j^i\right)^2+\left(\left.\frac{\partial\xi_2}{\partial s}\right|_j^i\right)+\left(\left.\frac{\partial\eta_2}{\partial s}\right|_j^i\right)
\end{aligned}
\right)\\
&+\left(\left.\frac{\partial\eta_2}{\partial s}\right|_j^i\right)\left(4\left.\frac{\partial\xi_2}{\partial s}\right|_j^i+2\left(\left.\frac{\partial\xi_2}{\partial s}\right|_j^i\right)^2+\frac{a^2}{2}\left(\left.\frac{\partial^2\xi_2}{\partial s^2}\right|_j^i\right)^2\right)
\end{aligned}
\right)\\
&+\frac{a^2}{2}K_c\left[
\begin{aligned}
&\left(\left.\frac{\partial\eta_2}{\partial s}\right|_j^i\right)^2\left(2\left(\left.\frac{\partial\xi_2}{\partial s}\right|_j^i\right)^2+\frac{a^2}{2}\left(\left.\frac{\partial^2\xi_2}{\partial s^2}\right|_j^i\right)^2\right)+\frac{a^2}{2}\left(\left.\frac{\partial^2\eta_2}{\partial s^2}\right|_j^i\right)^2\\
&-a^2\left(\left.\frac{\partial^2\eta_2}{\partial s^2}\right|_j^i\right)\left(\left.\frac{\partial\eta_2}{\partial s}\right|_j^i\right)\left(\left.\frac{\partial^2\xi_2}{\partial s^2}\right|_j^i\right)
\end{aligned}
\right]
\end{aligned}
\right] \qquad [A.6]
$$

The total energy of the pantograph unit cell writes accordingly:

$$W = W_1 + W_2 \qquad \text{[A.7]}$$

For the dynamical nonlinear analysis, all we need from the expression of the deformation energy is the quantities involving the third power of the strain; therefore, the deformation energy is written as:

$$W = \left[\begin{array}{l} \dfrac{a^2}{2} K_l \left(\begin{array}{l} +2\left(\begin{array}{l} \left(\left.\dfrac{\partial \xi_1}{\partial s}\right|_j^i \right)^3 + \left(\left.\dfrac{\partial \xi_2}{\partial s}\right|_j^i \right)^3 + \left(\left.\dfrac{\partial \eta_1}{\partial s}\right|_j^i \right)^3 + \left(\left.\dfrac{\partial \eta_2}{\partial s}\right|_j^i \right)^3 \\ + \left(\left.\dfrac{\partial \xi_1}{\partial s}\right|_j^i \right)^2 + \left(\left.\dfrac{\partial \xi_2}{\partial s}\right|_j^i \right)^2 + \left(\left.\dfrac{\partial \eta_1}{\partial s}\right|_j^i \right)^2 + \left(\left.\dfrac{\partial \eta_2}{\partial s}\right|_j^i \right)^2 \end{array} \right) \\ + \dfrac{a^2}{2} \left(\left.\dfrac{\partial^2 \xi_2}{\partial s^2}\right|_j^i \right)^2 + \dfrac{a^2}{2} \left(\left.\dfrac{\partial^2 \xi_1}{\partial s^2}\right|_j^i \right)^2 \\ + \left(\left.\dfrac{\partial \eta_1}{\partial s}\right|_j^i \right)^2 \left(2 \left.\dfrac{\partial \xi_1}{\partial s}\right|_j^i \right) + \left(\left.\dfrac{\partial \eta_2}{\partial s}\right|_j^i \right)^2 \left(2 \left.\dfrac{\partial \xi_2}{\partial s}\right|_j^i \right) \\ + \left(\left.\dfrac{\partial \eta_1}{\partial s}\right|_j^i \right) \left(2 \left(\left.\dfrac{\partial \xi_1}{\partial s}\right|_j^i \right)^2 \right) + \left(\left.\dfrac{\partial \eta_2}{\partial s}\right|_j^i \right) \left(2 \left(\left.\dfrac{\partial \xi_2}{\partial s}\right|_j^i \right)^2 \right) \end{array} \right) \\ + \dfrac{a^4}{4} K_c \left[\left(\left.\dfrac{\partial^2 \eta_1}{\partial s^2}\right|_j^i \right)^2 + \left(\left.\dfrac{\partial^2 \eta_2}{\partial s^2}\right|_j^i \right)^2 \right] \end{array} \right] \qquad \text{[A.8]}$$

We can easily extract the energy density, normalizing the unit cell energy by the unit cell area $S=\left(\sqrt{2}a\right)^2$

$$W/S=\left[\begin{array}{l}\frac{1}{4}K_l\left(\begin{array}{l}+2\left(\begin{array}{l}\left(\frac{\partial\xi_1}{\partial s}\bigg|_j^i\right)^3+\left(\frac{\partial\xi_2}{\partial s}\bigg|_j^i\right)^3+\left(\frac{\partial\eta_1}{\partial s}\bigg|_j^i\right)^3+\left(\frac{\partial\eta_2}{\partial s}\bigg|_j^i\right)^3\\+\left(\frac{\partial\xi_1}{\partial s}\bigg|_j^i\right)^2+\left(\frac{\partial\xi_2}{\partial s}\bigg|_j^i\right)^2+\left(\frac{\partial\eta_1}{\partial s}\bigg|_j^i\right)^2+\left(\frac{\partial\eta_2}{\partial s}\bigg|_j^i\right)^2\end{array}\right)\\+\frac{a^2}{2}\left(\frac{\partial^2\xi_2}{\partial s^2}\bigg|_j^i\right)^2+\frac{a^2}{2}\left(\frac{\partial^2\xi_1}{\partial s^2}\bigg|_j^i\right)^2\\+\left(\frac{\partial\eta_1}{\partial s}\bigg|_j^i\right)^2\left(2\frac{\partial\xi_1}{\partial s}\bigg|_j^i\right)+\left(\frac{\partial\eta_2}{\partial s}\bigg|_j^i\right)^2\left(2\frac{\partial\xi_2}{\partial s}\bigg|_j^i\right)\\+\left(\frac{\partial\eta_1}{\partial s}\bigg|_j^i\right)\left(2\left(\frac{\partial\xi_1}{\partial s}\bigg|_j^i\right)^2\right)+\left(\frac{\partial\eta_2}{\partial s}\bigg|_j^i\right)\left(2\left(\frac{\partial\xi_2}{\partial s}\bigg|_j^i\right)^2\right)\end{array}\right)\\+\frac{a^2}{8}K_c\left[\left(\frac{\partial^2\eta_1}{\partial s^2}\bigg|_j^i\right)^2+\left(\frac{\partial^2\eta_2}{\partial s^2}\bigg|_j^i\right)^2\right]\end{array}\right]\quad\text{[A.9]}$$

The transition is next made from the local basis (attached to the beam) to the Cartesian basis: the local components $\xi_{1j}^{\,i}$, $\xi_{2j}^{\,i}$, $\eta_{1j}^{\,i}$ and $\eta_{2j}^{\,i}$ project along the two components $\begin{bmatrix}u\\v\end{bmatrix}$. After development of equation [3.8], we obtain the energy density (for more details, see [RAH 15]).

3.6. Bibliography

[AND 14] ANDRIANOV I.V., DANISHEVS'KYY V.V., RYZHKOV O.I. *et al.*, "Numerical study of formation of solitary strain waves in a nonlinear elastic layered composite material", *Wave Motion*, vol. 51, no. 3, pp. 405–417, 2014.

[ASH 00] ASHBY M.F., EVANS A.G., FLECK N.A. *et al.*, *Metal Foams: A Design Guide*, Butterworth–Heinemann, Oxford, 2000.

[BAR 94] BARDENHAGSEN S., TRIANTAFYLLIDIS N., "Derivation of higher order gradient continuum theories in 2, 3-D non-linear elasticity from periodic lattice models", *Journal of the Mechanics and Physics of Solids*, vol. 42, pp. 111–139, 1994.

[BAT 79] BATHE K.J., BOLOURCHI S., *Large Displacement Analysis of Three Dimensional Beam Structures*, vol. 14, John Wiley & Sons, 1979.

[BHA 79] BHATNAGAR P.L., *Nonlinear Waves in One-dimensional Dispersive Systems*, vol. 142, Clarendon Press Oxford, 1979.

[BOU 72] BOUSSINESQ J., "Théorie des ondes et des remous qui se propagent le long d'un canal rectangulaire horizontal, en communiquant au liquide contenu dans ce canal des vitesses sensiblement pareilles de la surface au fond", *Journal de Mathématiques Pures et Appliquées, Deuxième Série*, vol. 17, pp. 55–108, 1872.

[CHR 07] CHRISTOV C., MAUGIN G., PORUBOV A., "On Boussinesq's paradigm in nonlinear wave propagation", *Comptes Rendus Mécanique*, vol. 335, pp. 521–535, 2007.

[DAR 06] DARAIO C., NESTERENKO V., HERBOLD E. *et al.*, "Of solitary wave properties in one-dimensional strongly nonlinear phononic crystals", *Physical Review E*, vol.73, 026610, 2006.

[DOS 12] DOS REIS F., GANGHOFFER J.F., "Equivalent mechanical properties of auxetic lattices from discrete homogenization", *Computational Materials Science*, vol.51, no. 1, pp. 314–321, 2012.

[ENG 11] ENGELBRECHT J., SALUPERE A., TAMM K., "Waves in microstructured solids and the Boussinesq paradigm", *Wave Motion*, vol. 48, pp. 717–726, 2011.

[GOF 02] GOFFAUX C., SANCHEZ-DEHESA J., YEYATI L. *et al.*, "Evidence of fano-like interference phenomena in locally resonant materials", *Physical Review B*, vol. 88, p. 225502, 2002.

[JAN 03] JANUS-MICHALSKA M., PECHERSKI R.P., "Macroscopic properties of open-cell foams based on micromechanical modeling", *Technische Mechanik*, vol. 23, pp. 221–231, 2003.

[JAN 05] JANUS-MICHALSKA M., "Effective models describing elastic behavior of cellular materials", *Archives of Metallurgy and Materials*, vol. 50, pp. 595–608, 2005.

[JAN 11] JANUS-MICHALSKA M., "Hyperelastic behavior of cellular structures based on micromechanical modeling at small strain", *Archives of Mechanics*, vol. 63, pp. 3–23, 2011.

[LIU 00] LIU Z., ZHANG X., MAO Y. *et al.*, "Locally resonant sonic materials", *Science*, vol. 289, pp. 1734–1736, 2000.

[LUR 90] LURE A.I., *Nonlinear Theory of Elasticity*, North-Holland, Amsterdam, 1990.

[MAN 13a] MANKTELOW K., NARISETTI RAJ K., LEAMY MICHEAL J. *et al.*, "Finite-element based perturbation analysis of wave propagation in nonlinear periodic structures", *Mechanical Systems and Signal Processing*, vol. 39, pp. 32–46, 2013.

[MAN 13b] MANKTELOW K., LEAMY MICHEAL J., RUZZENE M., "Topology design and optimization of nonlinear periodic materials", *Journal of the Mechanics and Physics of Solids*, vol. 61, pp. 2433–2453, 2013.

[NOR 99] NORRIS A.N., "Finite amplitude waves in solids", in HAMILTON M.F., BLACKSTOCK D.T. (eds), *Nonlinear Acoustics*, Academic Press, San Diego, 1999.

[OGD 97] OGDEN R.W., *Non-linear Elastic Deformations*, Courier Dover Publications, Dover, New York. 1997.

[PAR 06] PARNELL WILLIAM J., "Effective wave propagation in a pre-stressed nonlinear elastic composite bar", *Journal of Applied Mathematics*, vol. 72, pp. 223–244, 2006.

[POR 03] PORUBOV A., *Amplification of Nonlinear Strain Waves in Solids*, World Scientific, 2003.

[RAH 15] RAHALI Y., GIOROGIO I., GANGHOFFER J.F. *et al.*, "Homogenization à la Piola produces second gradient continuum models for linear pantographic lattices", *International Journal of Engineering Science*, vol. 97, pp. 148–172, 2015.

[RED 16a] REDA H., RAHALI Y., GANGHOFFER J.F. *et al.*, "Nonlinear dynamical analysis of 3D textiles based on second order gradient homogenized media", *Composite structures*, vol. 154, pp. 538–555, 2016.

[RED 16b] REDA H., RAHALI Y., GANGHOFFER J.F. *et al.*, "Wave propagation analysis in 2D nonlinear hexagonal periodic networks based on second order gradient nonlinear constitutive models", *International Journal of Non-Linear Mechanics*, vol. 87, pp. 85–96, 2016.

[RED 16c] REDA H., RAHALI Y., GANGHOFFER J.F. *et al.*, "Wave propagation in 3D viscoelastic auxetic and textile materials by homogenized continuum micropolar models", *Composite Structures*, vol. 141, pp. 328–345, 2016.

[RED 16d] REDA H., RAHALI Y., GANGHOFFER J.F. *et al.*, "Analysis of dispersive waves in repetitive lattices based on homogenized second-gradient continuum models", *Composite Structures*, vol. 152, pp. 712–728, 2016.

[RED 17] REDA H., GANGHOFFER J.F., LAKISS H., "Micropolar dissipative models for the analysis of 2D dispersive waves in periodic lattices", *Journal of Sound and Vibration*, vol. 392, pp. 325–345, 2017.

[SEN 08] SEN S., HONG J., BANG J. *et al.*, "Solitary waves in the granular chain", *Physics Reports*, vol. 46, pp. 21–66, 2008.

[SPA 07] SPADONI A., RUZZENE M., "Static aeroelastic response of chiral-core airfoils", *Journal of Intelligent Material Systems and Structures*, vol. 18, pp. 1067–1075, 2007.

[VAK 94] VAKAKIS A., KING M., PEARLSTEIN A., "Forced localization in a periodic chain of non-linear oscillators", *International Journal of Non-Linear Mechanics*, vol. 29, pp. 429–447, 1994.

[VIG 14] VIGLIOTTI A., DESHPANDE V.S., PASINI D., "Nonlinear constitutive models for lattice materials", *Journal of the Mechanics and Physics of Solids*, vol. 64, pp. 44–60, 2014.

[WAN 07] WANG J., NAUSIEDA A., LUCATO S. *et al.*, "Twisting of a high authority morphing structure", *International Journal of Solids and Structures*, vol. 44, no. 9, pp. 3076–3099, 2007.

[WAN 14] WANG P., CASADEI F., SHAN S. *et al.*, "Harnessing buckling to design tunable locally resonant acoustic metamaterials", *Physical Review Letters*, vol. 113, p. 014301, 2014.

[WAR 89] WARREN W.E., KRAYNIK A.M., STONE C.M., "A constitutive model for two-dimensional nonlinear elastic foams", *Journal of the Mechanics and Physics of Solids*, vol. 37, pp. 717–733, 1989.

[WAR 91] WARREN W.E., KRAYNIK A.M., "The nonlinear elastic behavior of open-cell foams", *Transactions of ASME*, vol. 58, pp. 375–381, 1991.

4

Multiscale Model of Concrete Failure

4.1. Introduction

The key assumption explored in this chapter pertains to the role of material heterogeneity in the study of the mechanical behavior of concrete. When a concrete specimen under extreme loading conditions enters a stage where cracks appear, the most appropriate model changes from a continuous into a discrete one. Crack propagation and final crack patterns might be based on accumulated microcracks of various types of initial flaws, defects in aggregate structure, voids or pores inside concrete caused during fabrication. This process of accumulation of microcracks leads to a complete failure mechanism, which is a brittle-type failure characteristic of concrete. In order to provide a reliable predictive model for failure of such heterogeneous material, we ought to find the solution at fine scale, where pre-existing microcracks continue to form into a large one during the increased loading leading to failure. The approach we focus upon in this chapter relies on spatial beam models, as a class of discrete lattice models [NIK 17], geometrically built using Delaunay triangulation [NIK 15a]. Here, the Delaunay edges can be considered as lattice elements, and their geometrical properties can be extracted from Voronoi cells (Figure 4.1).

Chapter written by Emir KARAVELIĆ, Mijo NIKOLIĆ and Adnan IBRAHIMBEGOVIĆ.

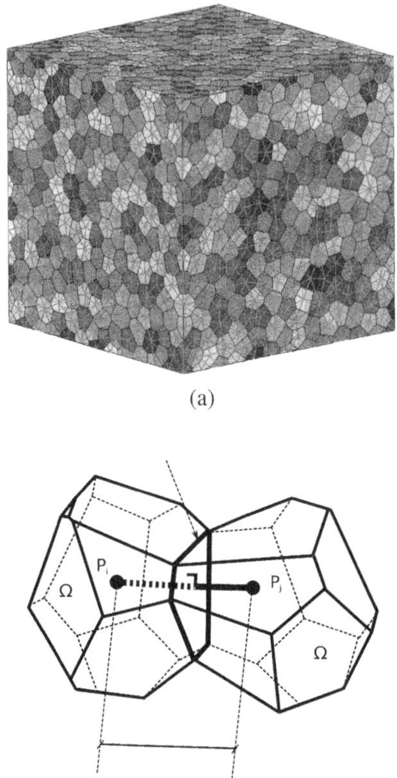

(a)

(b)

Figure 4.1. *Voronoi cells: (a) with 3D irregular lattice, (b) two neighborings. For a color version of the figure, see www.iste.co.uk/brancherie/microstructure.zip*

The main difficulty in failure analysis is to provide a correct, mesh-independent representation of the post-peak softening behavior [NIK 15b]. One of the ways to provide a mesh-independent response is by using the embedded strong discontinuity approach, which is described in this chapter [IBR 09, IBR 07]. The main reason for this mesh independency is that discontinuity or displacement jump always remains localized inside the element. This approach can be interpreted as a localization limiter that enhances the classical continuum mechanics theoretical formulation by admitting discontinuities in the displacement field. The numerical implementation of the discontinuity requires a modification of the standard finite-element procedure, which is similar to the method of incompatible modes [IBR 09, WAS 82, IBR 91].

Concrete is a highly non-homogeneous composite material. The model that can capture these heterogeneities should account for two or three different phases at least. Macroscopic mechanical response of such a material model will obviously depend on each phase's properties and their spatial distribution. Spatial distribution of each phase is based on a random process with Poisson distribution of the aggregates (phase I) and then filling the voids between them with cement paste (phase II). Separate study of each phase is not necessary because the formation of the cement layer can be determined on the basis of distribution of the aggregates. The distribution of the aggregates is taken into account as one of the two well-known grading curves for aggregates application in concrete mixture [NEV 10]. All particles with size less than 2.00 mm are included in the cement matrix volume, thus forming cement mortar. By using this way of distribution, we are able to obtain the volume fraction of each phase. The Gaussian distribution of the heterogeneity in each phase, aggregate and cement paste is taken into account, with standard deviation of parameter distribution restricted to $\pm\, 2\sigma$.

Various types of phenomenological constitutive models are widely used in structural scale computations for prediction of behavior in static or dynamic cases [LUB 90, LEM 04]. In the case of a complex loading program, these models require the proper criterion choice based on the studied material and the applied loading. Due to their macroscopic viewpoint, these models encounter difficulties in describing correctly the fine scale physical mechanisms such as fracture or damage. In this chapter, we present the meso-scale model capable of providing different failure mechanisms leading to macro-scale representation in terms of multi-surface models for concrete, combining Drucker–Prager for compression stress and Rankine for tensile stress. In order to provide such failure criteria, we carry out a large number of numerical tests at fine scale with different loading programs. We obtain compressive or tension strength as a function of aggregate volume fraction. The proposed model is also capable of describing the localized failure of structures [IBR 09].

This chapter is organized as follows: in section 2, we introduce the meso-scale model, with the cohesive links in terms of Timoshenko beam elements and a brief description of the model. In section 3, we present the corresponding macro-scale failure surface obtained with meso-scale model computations. Section 4 summarizes all the main findings and suggests the perspective study on this topic.

4.2. Meso-scale model

In this section, we give a brief description of the meso-model of a typical two-phase composite material, such as concrete. For more details, a complete description of the model, its numerical implementation and a number of illustrative examples of the model's predictive capabilities, the reader can refer to [NIK 17, NIK 15a, NIK 16]. In the presented model based on a 3D lattice finite-element model [SCH 92, SCH 97, YIP 05, BEN 10, OST 02], three failure modes are introduced through the kinematic enhancements: mode I representing tensile crack-mouth opening, mode II as shear sliding along the crack's mouth and mode III representing a tearing of material as a result of out-of-plane shear stress (Figure 4.2).

More precisely, enrichments in the displacement and strain fields are provided for translational directions x, y and z of the local coordinate system. The Timoshenko beams are used because of their ability to account for pronounced shear effects in both elastic and plastic phases. The shear contribution can be used to represent failure in modes II and III adding the corresponding displacement discontinuities. Thus, a set of complex failure patterns representing a complex crack evolution in 3 dimensions can be captured. The localization implies a heterogeneous strain field that no longer remains regular, even for a smooth stress field. The displacement field is written as the sum of smooth regular part $\bar{u}$ and the localized discontinuous part, which is represented as a product of the standard Heaviside function H_{xd} and the displacement jump $\tilde{u}$. The yield function used to activate the strong discontinuity is written as:

$$\Phi = t_\Gamma - (\sigma_u - q) \quad [4.1]$$

where t_Γ is the traction vector at the discontinuity, σ_u is the limit stress and q is a stress-like softening variable increasing by the exponential softening law:

$$q([|\alpha|]) = \sigma_u \left(1 - exp\left(-[|\alpha|]\frac{\sigma_f}{G_f}\right)\right) \quad [4.2]$$

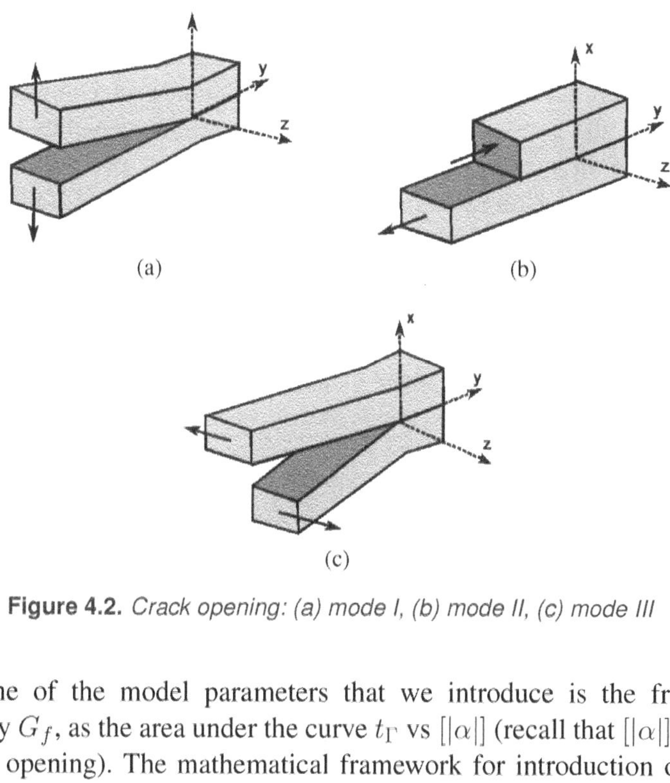

Figure 4.2. *Crack opening: (a) mode I, (b) mode II, (c) mode III*

One of the model parameters that we introduce is the fracture energy G_f, as the area under the curve t_Γ vs $[|\alpha|]$ (recall that $[|\alpha|]$ is the crack opening). The mathematical framework for introduction of this kind of enhancement is provided by the method of incompatible modes and Hu-Washizu mixed variational formulation, where the displacement field is separated from strain and stress virtual fields [WAS 82, IBR 91]. After discretization of the displacement field using the enhanced finite-element method (E-FEM) [IBR 07], the following system needs to be solved:

$$\mathbf{A}_{e=1}^{n_{el}} \left[\mathbf{f}_e^{int}(d, [|\alpha|]) - \mathbf{f}_e^{ext} \right] = 0 \qquad [4.3]$$

$$t_\Gamma - (\sigma_u - q) = 0$$

where $\mathbf{A}$ denotes the standard assembly operator [IBR 09, BAT 06, ZIE 00]. The computational procedure used to obtain the solution of equation [4.3] is explained in detail in [NIK 17, IBR 09].

In order to understand the failure behavior of concrete-like materials, it is necessary to study crack growth under mixed modes I, II and III in the presence of heterogeneities. Heterogeneities play a crucial role in making the computational iterative procedure more robust by eliminating the academic case of localized failure of homogeneous material under a homogeneous stress field. The computational model presented herein thus leads to a more robust iterative procedure. In order to provide this, we use a random process with Gaussian distribution to define limit stress for each phase with restriction to $\pm\, 2\sigma$, setting mean value and standard deviation for each limit stress (Figure 4.3).

Each phase is spread across the specimen by the Poisson distribution, taking into account the grading curves for aggregate [NEV 10]. Two well-known grading curves, Fuller and EMPA (Figure 4.4), are defined by the following relations:

$$p(d_F) = 100\sqrt{\frac{d}{d_{max}}} \qquad [4.4]$$

$$p(d_E) = 50\left(\frac{d}{d_{max}} + \sqrt{\frac{d}{d_{max}}}\right)$$

where $p(d_F)$ and $p(d_E)$ are the cumulative sieve fractions passing through a sieve with diameter d and d_{max} is the diameter of the coarsest aggregate.

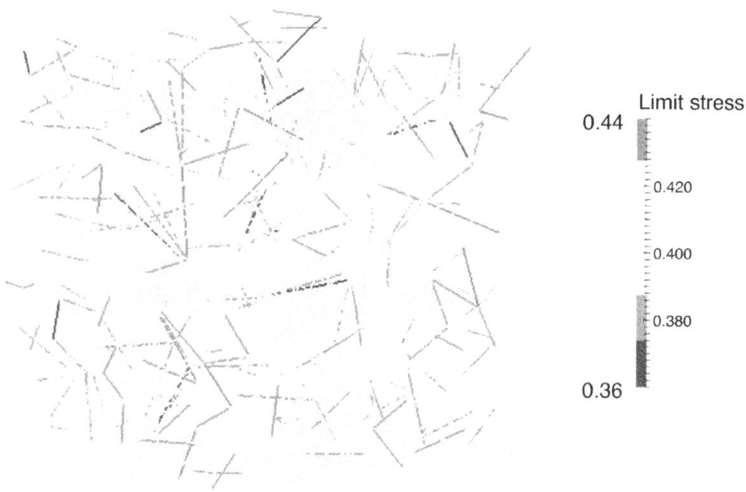

Figure 4.3. *Variation of input material parameters. For a color version of the figure, see www.iste.co.uk/brancherie/microstructure.zip*

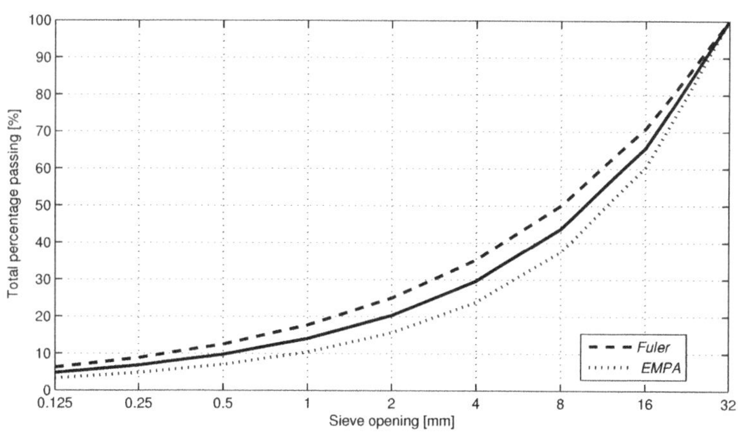

Figure 4.4. *Grading curves*

The distribution of the aggregate is taken into account as arithmetic mean of these two grading curves where all particles smaller than 2.00 mm are included in the cement matrix volume, thus forming cement mortar. A separate study of each phase is not necessary because the formation of the cement layer can be determined on the basis of the distribution of the aggregates. According to the defined grading curve we use in further study, the aggregate with maximal diameters of 8, 16 and 32 mm obtains to 30%, 50% and 60% volume fractions for phase I, respectively.

4.3. Macroscopic model response

In this section, we present the computed macroscopic responses for a number of different macroscopic loading paths. The computations are performed by a research version of computer program FEAP, developed by R.L. Taylor at UC Berkeley [TAY 17].

4.3.1. *Uniaxial tests*

4.3.1.1. *Construction of specimen*

As already elaborated in the previous section, the approach we present relies on models at meso-scale with cohesive links in terms of spatial beams. Such beams are generated by computing the 3D Delaunay triangulation, which is performed using Gmsh [GEU 09] over the spatial domain of interest. The edges of the resulting tetrahedral elements are converted into the beams whose cross-sections are computed from the corresponding Voronoi tessellation. Such task is computed using code written with MATLAB.

4.3.1.2. *Tension test*

For dealing with a uniaxial tension test, we consider the specimen given as a cube with 15 cm side length with a different volume fraction of aggregates (phase I). Table 4.1 summarizes the chosen mechanical properties for each phase. We can note that aggregates remain in the elastic regime, which is far more stiff than the cement matrix. This computation is made under the displacement control.

Cement matrix	Aggregate
E=15 GPa	E=70 GPa
ν=0.25	ν=0.25
σ_{ut}=4.0 MPa	-
σ_{uc}=40.0 MPa	-
$\tau_{u,v,w}$=1.0 MPa	-
G_{fu}=6 N/m	-
$G_{fv,w}$=100 N/m	-
Φ=20°	-

Table 4.1. *Uniaxial tension test: materials parameters*

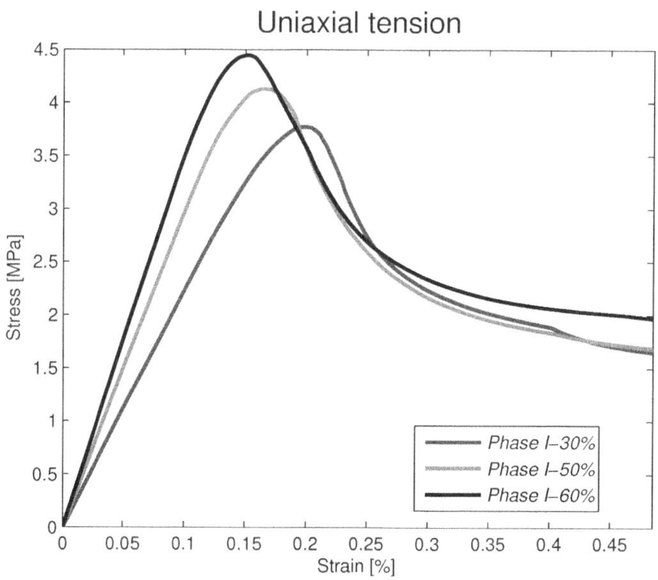

Figure 4.5. *Uniaxial tension test: macroscopic response. For a color version of the figure, see www.iste.co.uk/brancherie/microstructure.zip*

Figure 4.5 shows macroscopic stress (sum of all reactions in Z direction per cross-sectional area) versus strain curve. The macroscopic Young's modulus and limit stress trigger the softening changes due to the volume fraction of phase I for 30%, 50% and 60%.

With an increase of phase I volume fraction, the global modulus of elasticity as well as the elastic limit stress point increase. This is the consequence of using more elements of phase I, which makes the global response of the specimen less ductile. It can also be seen that when the volume fraction ratio of phase II increases, the failure of the specimen becomes more ductile, with a larger fracture process zone, and more brittle in the softening response. The latter is due to the appearance of many more potential macro-cracks, which drive the stress to zero more quickly. The main physical explanation of this influence lies in the total volume of voids in concrete. The structure of cement paste is complex and several sources of flaws and defects exist even before the application of external load; up to 50% of the volume of cement paste may consist of pores (gel pores, capillary pores and accidental or entrapped air). If we ignore all voids in aggregate (for normal aggregates, these are minimal) with increasing aggregate-to-cement ratio, the cement paste will represent a smaller proportion of specimen volume. Thus, the total porosity is lower and hence the limit stress point is higher [NEV 10].

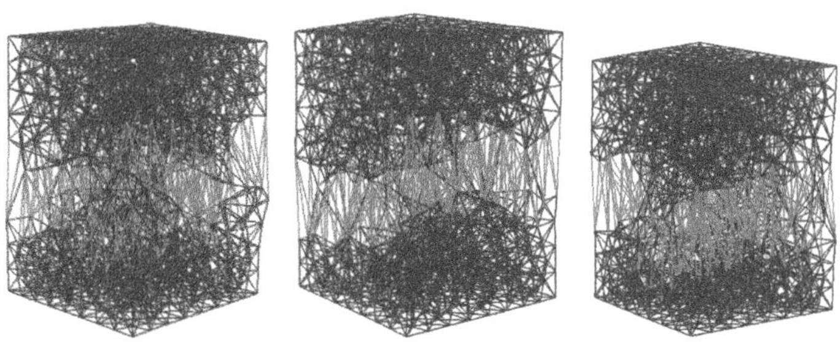

(a) (b) (c)

Figure 4.6. *Uniaxial tension test-deformed mesh with increasing softening elements in mode I: (a) phase I-30%; (b) phase I-50%; (c) phase I-60%*

Figure 4.6 presents beam elements in subsequent stages of increasing softening at the end of tension test computation for three different heterogeneous specimens. This can be clearly seen in Figure 4.6, in which a macro-crack lies in a direction perpendicular to the imposed displacement. One dominant macro-crack is present in any specimen, inducing the final failure mechanism and formed differently depending on the distribution of the weaker phase, which decides the final crack position. Failure due to mode I is more pronounced in tension tests.

4.3.1.3. *Compression test*

In this part, we present the results of numerical simulations of macroscopic response for specimens under a uniaxial compression loading program. Geometric, material parameters and the distribution of aggregate and cement paste are the same as for the specimen used under uniaxial tension test (see Table 4.1). Macroscopic response (Figure 4.7) reveals the changes in modulus of elasticity and elastic limit stress point due to different volume fractions of the aggregate. With an increase of phase I, the global modulus of elasticity and elastic limit point increase similarly to the tension test case. The difference concerns the ductile phase of the response during creation of the fracture process zone, which is more pronounced in the compression test than in the tension test. Thus, not only is the ductile part with fracture process zone larger, but also the contribution of compression failure. The main reason for this is that more elements are subjected to shear and compression simultaneously, where the crack propagates because of the shear, which leads to mode II or mode III failure. If the crack propagates in mode II or III, it is still possible to transfer the compression force through the specimen, assuming that the two separated blocks formed during cracking in mode II or III lean on each other. Compression force in this situation increases up to the point of significant damage on the specimen and until cracking is extensive enough that loading capacity starts to decrease.

Figure 4.8 presents beam elements in increasing softening at the end of compression test computations for three different heterogeneous specimens. Contrary to tension test crack patterns, in the compression test, many more macro-cracks are needed to drive the specimens to failure and these are influenced more by mode II and mode III mechanisms, which form the final crack patterns together with mode I.

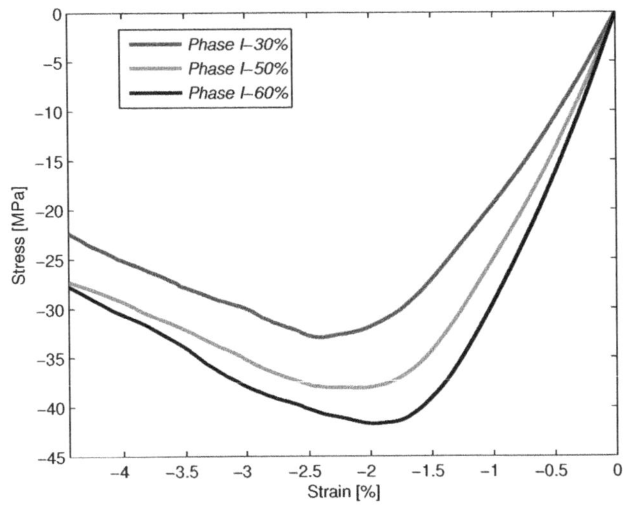

Figure 4.7. *Uniaxial tension test: macroscopic response. For a color version of the figure, see www.iste.co.uk/brancherie/microstructure.zip*

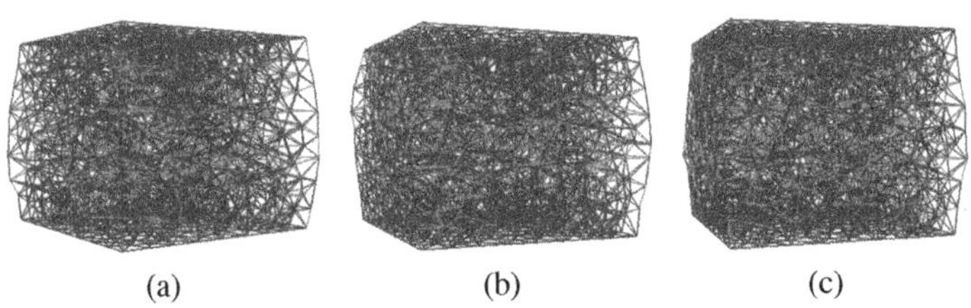

Figure 4.8. *Uniaxial compression test-deformed mesh with increasing softening elements in mode I: (a) phase I-30%; (b) phase I-50%; (c) phase I-60%. For a color version of the figure, see www.iste.co.uk/brancherie/microstructure.zip*

In order to compare the macroscopic responses corresponding to tension and compression, Figure 4.9 presents both macroscopic stress versus strain curves and reveals that the overall compression-to-tension ratio is equal to 8.8, 9.37 and 9.50 for 30%, 50% and 60% of phase I, respectively.

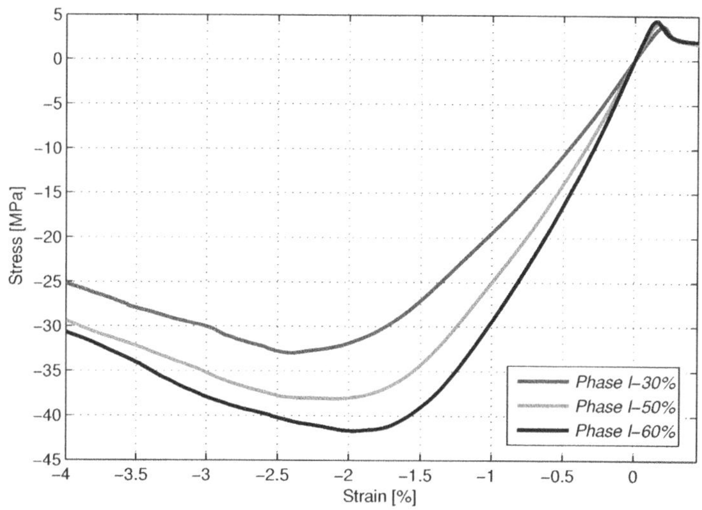

Figure 4.9. *Complete macroscopic response. For a color version of the figure, see www.iste.co.uk/brancherie/microstructure.zip*

4.3.2. *Failure surface*

In this section, we show how to obtain multi-surface criteria for concrete-like materials by taking into account heterogeneities and different processes of cracking using the meso-scale model proposed in the previous section. The most appropriate combination of multi-surface models for concrete can be considered combining the Drucker–Prager model for compression stress and Rankine model for tensile stress, with a damage model describing the localized failure of structure. It can be written in terms of the principal stress values according to:

$$\begin{aligned} \Phi^{DP} &= \sqrt{J_2} + \mu I_1/\sqrt{6} - (f_c - \hat{q}^c(\xi^c)) \leq 0 \\ \Phi_i^R &= \sigma_i - (f_t - \hat{q}^t(\xi^t)) \leq 0 \end{aligned} \quad [4.5]$$

where Φ^{DP} and Φ_i^R are, respectively, the Drucker–Prager and Rankine yield surfaces; σ_i, J_2 and I_1 are principal values, the second and the first invariants of stress tensor; f_c and f_t are compressive and tensile yield stress values and q^c and q^t are the corresponding internal variables that control the plasticity threshold evolution resulting in a desired amount of fracture energy for any particular mode of localized failure [BEN 12].

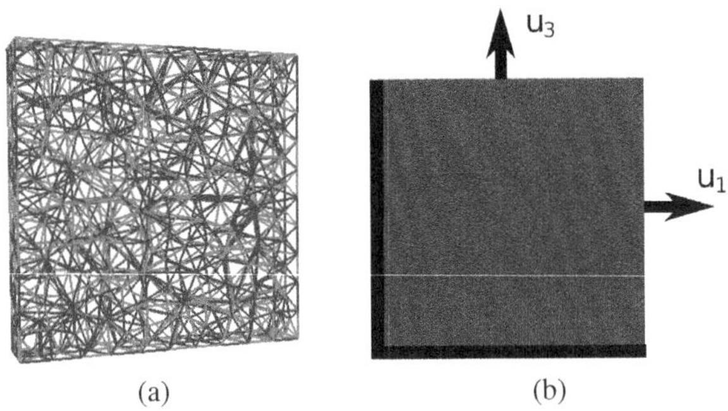

Figure 4.10. *Specimen: (a) with 50% of aggregates (red-colored elements); (b) boundary conditions and loading. For a color version of the figure, see www.iste.co.uk/brancherie/microstructure.zip*

In order to test biaxial failure behavior, we choose the plate specimen with dimensions of 150 mm x 3 mm x 150 mm (Figure 4.10). The plate is subjected to biaxial imposed displacements, producing the following stress combinations: tension–tension, compression–compression, compression–tension and tension–compression. Uniaxial compressive strength value is taken from a uniaxial compressive test performed on a plate concrete specimen (Table 4.2).

Direction i	Tension strength σ_{ti} (MPa)	Compressive strength σ_{ci} (MPa)
$1(X)$	2.744	–26.623
$3(Z)$	2.636	–23.266

Table 4.2. *Uniaxial strength with 50% of aggregates*

In order to obtain these four combinations (compression–compression, tension–tension, compression–tension and tension–compression), we perform a circular loading program on the basis of the trigonometrical circle $cos(\alpha)^2 + sin(\alpha)^2 = 1$. Thus, a couple of imposed displacements ($u_1 = cos(\alpha)$, $u_3 = sin(\alpha)$) are parametrized in terms of angle α. Faces X = 150 mm and Z = 150 mm are, respectively, subjected to imposed displacements u_1 and u_3, whereas faces X = 0 mm and Z = 0 mm are blocked in X and Z directions, respectively (Figure 4.10(b)). Compression and tension are chosen as negative and positive values, respectively. Stress in the principal direction 1 (X direction) is labeled as $\overline{\sigma}_1$, and stress in principal direction 3 (Z direction) is labeled as $\overline{\sigma}_3$. The ultimate strength in the principal direction 1 is written as σ_1, and the one in direction 3 is denoted by σ_3. Table 4.3 presents the ratio of ultimate strengths σ_1 and σ_3 to σ_{ci} under different couples of imposed displacements. Graphical representation of Table 4.3 is presented in Figure 4.14, while Figure 4.15 shows the evolution of σ_1/σ_{ci} with respect to σ_3/σ_{ci} during loading until failure. Table 4.4 presents the computed macroscopic fracture energy under particular couples of imposed displacements. First, we observe the uniaxial test versus the biaxial test for both tension and compression (Figure 4.11). It can be noted that biaxial response is stiffer but less ductile compared to uniaxial response in either the X or Z direction and presents a higher ultimate strength than the uniaxial one. Figures 4.12 and 4.13 present beam elements in increasing softening at the end of both tension and compression test computations.

Symmetrical behavior of concrete in relation with the 45° line under biaxial loading is observed in Figure 4.14, as well as in experimental approaches [SAN 04]. As already mentioned, the ultimate compression strength of concrete is higher under the biaxial compression test than in the uniaxial one. Physically, this is due to the decreasing of porosity for biaxial compression. An increase in ultimate compression strength in relation to the corresponding uniaxial compression strength is observed for each couple of imposed displacement in the compression–compression region. Maximum ultimate compression strength is obtained for couples of imposed displacements $u_1 = -0.1800$ mm, $u_3 = -0.0872$ mm and $u_1 = -0.0872$ mm, $u_3 = -0.1800$ mm for σ_1 and σ_3, respectively. In the tension–tension region, the value of the concrete ultimate strength for each couple of imposed displacement is very close to the corresponding uniaxial tensile strength. Finally, in tension–compression and compression–tension regions, concrete compression ultimate strength increases while the tension decreases.

Combined regions	u_1(mm),u_3(mm)	σ_1/σ_c	σ_3/σ_c
Tension–tension	0.0000,0.0200	0.000	–0.112
	0.0173,0.0100	–0.113	–0.091
	0.0100,0.0173	–0.097	–0.120
	0.0200,0.0000	–0.104	0.000
	0.0056,0.0192	–0.065	–0.119
	0.0192,0.0056	–0.112	–0.057
	0.0200,0.0200	–0.111	–0.117
Compression–compression	0.0000,–0.2000	0.000	1.000
	–0.1732,–0.1000	1.220	0.969
	–0.1000,–0.1732	1.040	1.251
	–0.2000,0.0000	1.000	0.000
	–0.0872,0.1800	0.976	1.248
	–0.1800,–0.0872	1.224	0.862
	–0.2000,–0.2000	1.181	1.186
	–0.1990,–0.0200	1.192	0.445
	–0.0200,–0.1990	0.421	1.152
	–0.1950,–0.0444	1.224	0.589
	–0.0444,–0.1950	0.647	1.208
	–0.2000,–0.0020	1.156	0.346
	–0.0020,–0.2000	0.315	1.126
Tension–compression	0.1732,–0.1000	–0.107	0.639
	0.1936,–0.0500	–0.106	0.512
	0.0224,–0.0380	–0.071	0.810
	0.0173,–0.0200	–0.082	0.707
Compression–tension	–0.1732,0.1000	0.604	–0.130
	–0.0100,0.1936	0.507	–0.124
	–0.0380,0.0224	0.801	–0.083
	–0.0200,0.0173	0.686	–0.100

Table 4.3. *Ratios of ultimate strengths to σ_c along X and Z directions under different couples of imposed displacements*

Combined regions	u_1 (mm), u_3 (mm)	G_f (N/m)
Tension–tension	0.0200,0.0200	6.525
Compression–compression	–0.2000,–0.2000	357.351
Tension–compression	0.0224,–0.0380	73.157
Compression–tension	–0.0380,0.0224	72.241

Table 4.4. *Fracture energy under particular couples of imposed displacements*

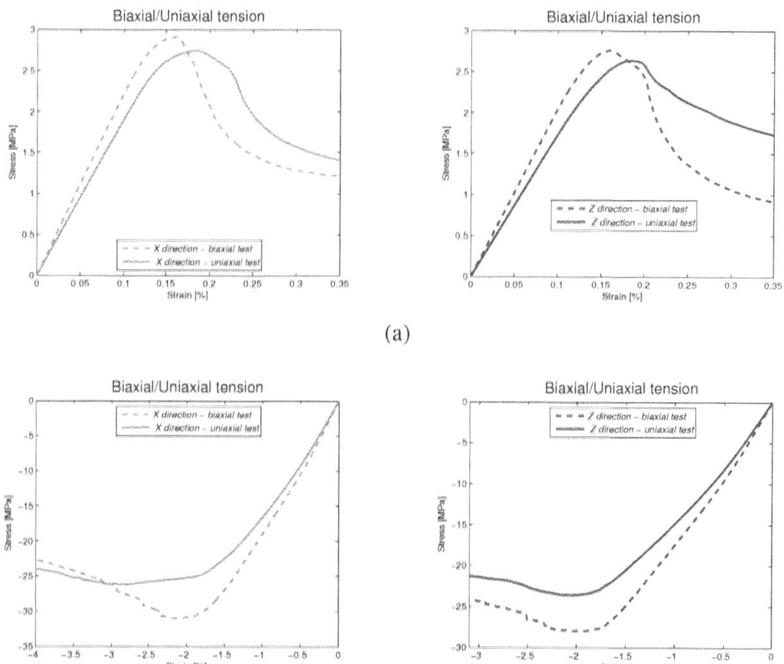

Figure 4.11. *Biaxial–uniaxial test comparison: (a) macroscopic response in X and Z directions – tension test; (b) macroscopic response in X and Z directions – compression test*

Figure 4.15 shows that in the compression–compression region, behavior of concrete is linear between σ_1/σ_{ci} and σ_3/σ_{ci} due to prevented crack opening. Moreover, biaxial compression loading in X and Z directions tends to prevent crack opening, respectively, in Z and X directions and the dominant failure mechanism occurs in the Y direction (out of plane). For other regions (tension–tension, tension–compression and compression–tension), a linear relation between σ_1/σ_{ci} and σ_3/σ_{ci} is first observed until cracking is not present, and a nonlinear relation between them is observed until cracking is more pronounced. In the tension–compression and compression–tension regions, we can note that the ultimate tension strength is reached sooner compared to compression due to a higher ratio of imposed displacements.

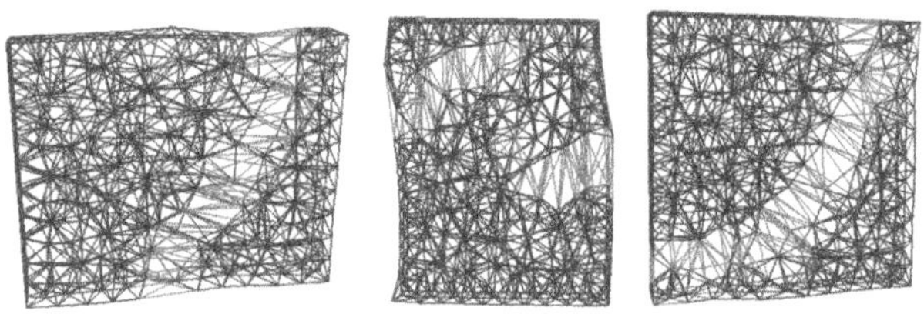

(a) (b) (c)

Figure 4.12. *Tension test-deformed mesh with increasing softening elements in mode I: (a) uniaxial test – X direction; (b) uniaxial test – Z direction; (c) biaxial test. For a color version of the figure, see www.iste.co.uk/brancherie/microstructure.zip*

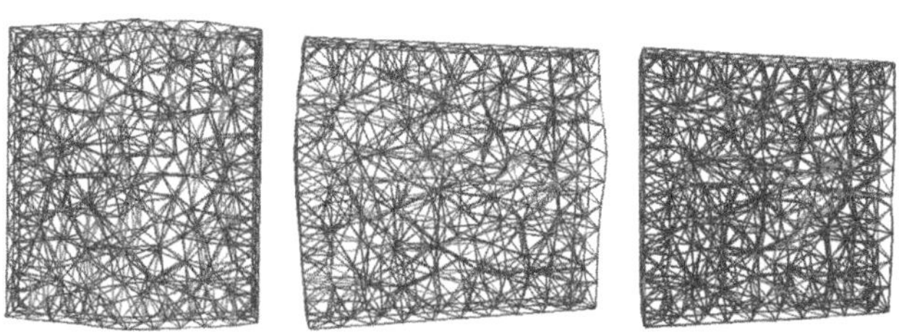

(a) (b) (c)

Figure 4.13. *Compression test-deformed mesh with increasing softening elements in mode I: (a) uniaxial test – X direction; (b) uniaxial test – Z direction; (c) biaxial test. For a color version of the figure, see www.iste.co.uk/brancherie/microstructure.zip*

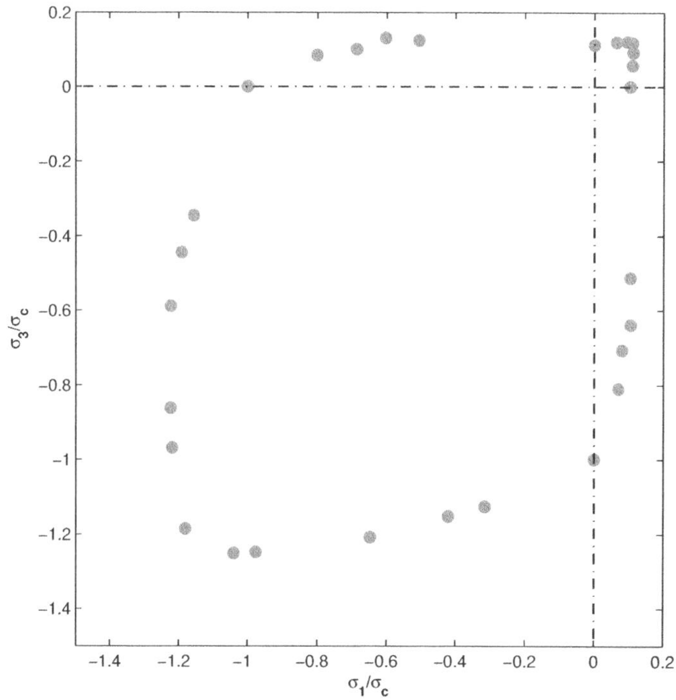

Figure 4.14. *Failure surface*

4.4. Conclusions

In this chapter, we proposed a novel three-dimensional beam lattice model with the ability to represent the failure phenomena for concrete-like materials. The model is able to provide the corresponding macro-scale failure criteria, combining the Drucker–Prager model for compression stress and the Rankine model for tensile stress. The main findings proposed in this chapter are obtained with numerical tests carried out with a meso-scale model, which relies on cohesive links as spatial beam elements representing cement paste and aggregates. Any such element is represented by 3D Timoshenko beams, embedded with strong discontinuities in local coordinate system directions, which provide the capability to simulate the localized failure in modes I, II and III. Failure in all three modes can not only occur individually but also be presented in mixed mode. The failure criterion is different under the force of tension or compression. Under tension, softening is activated simultaneously for all modes as soon as one failure surface becomes active, whereas under compression, the failure modes are handled separately. The compression force influences the shear strength threshold in a Mohr–Coulomb manner. Another feature of the model is that each material phase parameters are introduced through the Gaussian distribution representing heterogeneities of each phase. The model of this kind can be used successfully to present quite complex macroscopic responses.

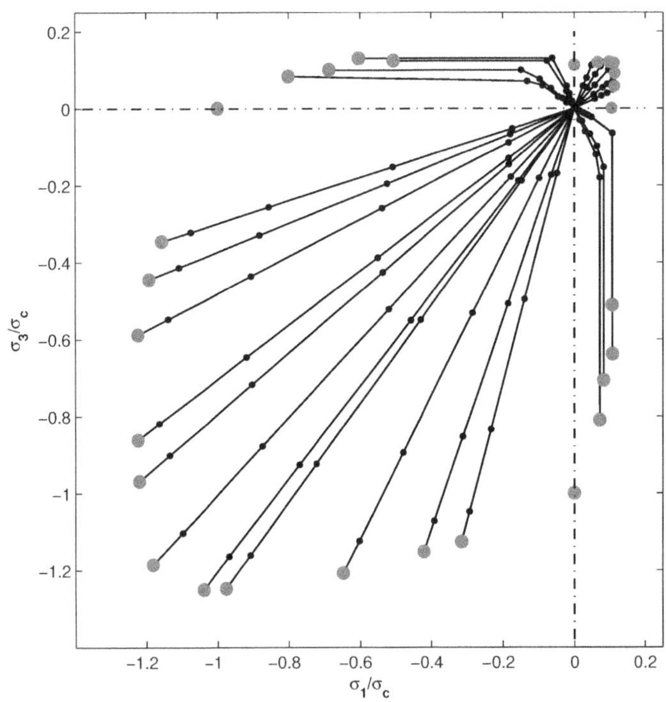

Figure 4.15. *Evolution of σ_1/σ_{ci} with respect to σ_3/σ_{ci} during loading until failure*

Three concrete specimens with 30%, 50% and 60% volume fractions of aggregate were subjected to uniaxial tension and compression tests. We found that the resistance to tension is much lower than that for compression. We observed one dominant macro-crack subjected to mode I that led the specimen to failure under tension, whereas under compression, many more macro-cracks were needed to drive the specimens to failure, with a significant contribution of mode II and mode III mechanisms. We also found that ultimate strength increases with the increasing volume fraction of aggregates and also influences the crack path. Overall compression–tension ratio σ_c/σ_t was equal to 8.8, 9.37 and 9.50 for 30%, 50% and 60% of phase I, respectively. All these points fit well with typical observations made for concrete-like materials [KUP 69, CAR 84, AST 02].

Finally, one of the most significant findings is that this approach can provide the definition of parameters such as the fracture energy and ultimate strength, taking into account the heterogeneity and different processes of cracking governing the final failure mechanism. This approach can also be used for parameter identification of multi-surface models (e.g. the combination of the Rankine and Drucker–Prager criteria [DOL 07]) providing more predictive results from classical phenomenological models in structural analysis.

This chapter opens up perspectives for studying the influence of the probability distribution of each phase on failure parameters.

4.5. Acknowledgments

Joint support by Haut-de-France Region (CR Picardie) (120-2015-RDISTRUCT-000010 and RDISTRUCT-000010) and EU funding (FEDER) for Chaire-de-Mecanique (120-2015-RDISTRUCTF-000010 and RDISTRUCTI-000004) and a scholarship from the French Embassy in Sarajevo are gratefully acknowledged.

4.6. Bibliography

[AST 02] ASTUDILLO DE LA VEGA E., "Etude expérimentale du comportement mécaniquedún Béton Fibré à Ultra Hautes Performances (BFUHP) en traction biaxiale", PhD Thesis, Ecole Normale Superieure de Cachan, 2002.

[BAT 06] BATHE K.J., *Finite Element Procedures*, Prentice Hall, Upper Saddle River, 2006.

[BEN 10] BENKEMOUN N., HAUTEFEUILLE M., COLLIAT J.B. *et al.*, "Failure of heterogeneous materials: 3D meso-scale FE models with embedded discontinuities", *International Journal for Numerical Methods in Engineering*, vol. 82, pp. 1671–1688, 2010.

[BEN 12] BENKEMOUN N., IBRAHIMBEGOVIC A., COLLIAT JB., "Anisotropic constitutive model of plasticity capable of accounting for details of meso-structure of two-phase composite material", *Computers & Structures*, vol. 90–91, pp. 153–162, 2012.

[CAR 84] CARPINTERI A., INGRAFFEA A.R. (eds), *Fracture Mechanics of Concrete*, Martinus Nijhoff, The Netherlands, 1984.

[DOL 07] DOLAREVIC S., IBRAHIMBEGOVIC A., "A modified three-surface elasto-plastic cap model and its numerical implementation", *Computers & Structures*, vol. 85, pp. 419–30, 2007.

[GEU 09] GEUZAINE C., REMACLE J.F., "Gmsh: a three-dimensional finite element mesh generator with built-in pre- and post-processing facilities", *International Journal for Numerical Methods in Engineering*, vol. 11, pp. 1309–1331, 2009.

[IBR 91] IBRAHIMBEGOVIC A., WILSON E., "A modified method of incompatible modes", *Communications in Numerical Methods in Engineering*, vol. 7, pp. 187–194, 1991.

[IBR 07] IBRAHIMBEGOVIC A., MELNYK S., "Embedded discontinuity finite element method for modeling of localized failure in heterogeneous materials with structured mesh: an alternative to extended finite element method", *Computational Mechanics*, vol. 40, pp. 149–155, 2007.

[IBR 09] IBRAHIMBEGOVIC A., *Nonlinear Solid Mechanics: Theoretical Formulations and Finite Element Solution Methods*, Springer, London, 2009.

[KUP 69] KUPFER H., HILSDORF H.K., "Behavior of concrete under biaxial stress", *ACI Journal Proceedings*, vol. 66, no. 8, pp. 656–666, 1969.

[LEM 04] LEMAITRE J., CHABOCHE J.-L., *Mécanique des Matériaux Solides*, Dunod, Paris, 2004.

[LUB 90] LUBLINER J., *Plasticity Theory*, Macmillan, London, 1990.

[NEV 10] NEVILLE A.M., BROOKS J.J., *Concrete Technology*, 2nd edition, Prentice Hall, Upper Saddle River, 2010.

[NIK 15a] NIKOLIC M., IBRAHIMBEGOVIC A., "Rock mechanics model capable of representing initial heterogeneities and full set of 3D failure mechanisms", *Computer Methods in Applied Mechanics and Engineering*, vol. 290, pp. 209–227, 2015.

[NIK 15b] NIKOLIC M., IBRAHIMBEGOVIC A., MISCEVIC P., "Brittle and ductile failure of rocks: embedded discontinuity approach for representing mode I and mode II failure mechanisms", *International Journal for Numerical Methods in Engineering*, vol. 102, pp. 1507–1526, 2015.

[NIK 16] NIKOLIC M., IBRAHIMBEGOVIC A., MISCEVIC P., "Discrete element model for the analysis of fluid-saturated fractured poro-plastic medium based on sharp crack representation with embedded strong discontinuities", *Computer Methods in Applied Mechanics and Engineering*, vol. 298, pp. 407–427, 2016.

[NIK 17] NIKOLIC M., KARAVELIC E., IBRAHIMBEGOVIC A. *et al.*, "Lattice element models and their peculiarities", *Archives of Computational Methods in Engineering*, pp. 1–32, doi: 10.1007/s11831-017-9210-y, 2017.

[OST 02] OSTOJA-STARZEWSKI M., "Lattice models in micromechanics", *Applied Mechanics Reviews* vol. 55, no. 1, pp. 35–60, 2002.

[SAN 04] SANG-KEUN L., YOUNG-CHUL S., SANG-HOON H., "Biaxial behavior of plain concrete of nuclear containment building", *Nuclear Engineering and Design*, vol. 227, pp. 143–53, 2004.

[SCH 92] SCHLANGEN E., VAN MIER J.G.M., "Simple lattice model for numerical simulation of fracture of concrete materials and structures", *Materials and Structures*, vol. 25, pp. 534–542, 1992.

[SCH 97] SCHLANGEN E., GARBOCZI EJ., "Fracture simulations of concrete using lattice models: computational aspects", *Engineering Fracture Mechanics*, vol. 57, pp. 319–332, 1997.

[TAY 17] TAYLOR R.L., FEAP finite element analysis program, University of California, Berkeley, available at: http://www.ce.berkeley.edu.rlt, 2017.

[WAS 82] WASHIZU K., *Variational Methods in Elasticity and Plasticity*, 3rd edition, Pergamon Press, New York, 1982.

[YIP 05] YIP M., MOHLE J., BOLANDER J.E., "Automated modeling of three-dimensional structural components using irregular lattices", *Computer-Aided Civil and Infrastructure Engineering*, vol. 120, pp. 393–407, 2005.

[ZIE 00] ZIENKIEWICZ O.C., TAYLOR R.L., *Finite Element Method*, 5th edition, Butterworth-Heinemann, Oxford, 2000.

5

Discrete Numerical Simulations of the Strength and Microstructure Evolution During Compaction of Layered Granular Solids

5.1. Introduction

Compaction of granular materials and powders is a very common practice in several engineering and manufacturing applications [HEC 61]. Powders can be compacted using roller compaction, where powders are forced through a pair of rotating rollers. Powders can also be compacted using direct compaction, where powders inside a die are compacted in the axial direction. One of the main purposes of powder compaction is to form solids such as pharmaceutical tablets and direct compaction is mostly the preferred method to form the tablets. A lot has been learned from prior experimental studies on tablet compaction, including relationship between porosity and compaction pressure [HEC 61], the effects of particle size on powder compaction [ADO 97, ALM 08, EIC 09, FIC 05, HER 07, KHO 13, MCK 82, PAT 07, SUN 06, SUN 08, YOH 15], the role of lubricants on compaction and tensile strength of tablets [ALM 08, SUN 06], to mention only a few.

Chapter written by Bereket YOHANNES, Marcial GONZALEZ and Alberto M. CUITIÑO.

The above-mentioned studies focus only on monolayer tablets, in which the entire power bed is exposed to the same applied compaction pressure. In recent years, there was some progress in the areas of layered powder compacts, especially for bilayer pharmaceutical tablets [ABE 14]. Bilayer tablets are useful for controlled drug delivery of two different active ingredients [ABE 14, ABD 04], and separation of chemically incompatible drugs [DIV 11]. In layered powder compaction, layers are compacted in a sequential manner. For example, compaction of a bilayer tablet consists of the following steps:

1) Filling die with the first-layer particles;

2) Compacting the first-layer particles (i.e. application of the first-layer compaction force);

3) Removing the first-layer compaction force;

4) Filling die with the second-layer particles on top of the first-layer particles;

5) Compacting the two layers together to form the bilayer tablet (i.e. application of the second-layer compaction force);

6) Removing the second-layer compaction force;

7) Ejecting the tablet from the die.

If more layers are needed, the process is repeated without ejecting the compacted layers from the die.

The production of these layered (both bilayer and multilayered) tablets poses a challenge as the layered tablets are prone to fracture by delamination, usually along the interfaces between two layers, because of their inherent insufficient strength [AKS 13, KOT 12]. Thus, one of the main manufacturing challenges is to obtain tablets that do not fracture at the interface because of insufficient mechanical strength.

The mechanical strength of a bilayer tablet depends on many factors. Several researchers have experimentally investigated the factors that influence the mechanical strength of these tablets, particularly the mechanical strength at the interface of the layers.

Using axial tensile strength tester, Akseli *et al.* [AKS 13] experimentally studied the strength of bilayer tablets made up of microcrystalline cellulose (MCC) and starch powders (Figure 5.1(a)). They studied the effect of the first- and second-layer compaction forces on the mechanical strength of the tablets and types of failures associated with it. They showed that the tensile strength of the bilayer tablets increased as the second-layer compaction force is increased. The increase in the tensile strength with the compaction force is also true for monolayer tablets [ADO 97, YOH 15, YOH 16]. The most interesting finding from experiments conducted by Akseli *et al.* [AKS 13] is that the tensile strength of the tablets does not increase monotonically with the magnitude of the first-layer compaction force, when the second-layer force is kept constant. For lower first-layer compaction forces, the tensile strength increases as the first-layer compaction force is increased. However, at higher first-layer compaction forces, the tensile strength decreases as the first-layer compaction force is increased. The highest tensile strength was achieved when the first-layer compaction force was 4 kN for a 10 mm-diameter tablet of both MCC–MCC and MCC–starch bilayers. (MCC–MCC refers to a tablet where both the first and second layers are composed of MCC powder, while MCC–starch refers to a tablet where the first layer is MCC powder and the second layer is starch powder).

Kottala *et al.* [KOT 12] showed that the layer interface surface roughness also significantly influences the mechanical properties of the bilayer tablet. They found that high interfacial roughness and the interfacial curvature increase bonding between adjacent layers, which tend to increase the interfacial strength of the tablets. Both the interfacial surface roughness and interfacial curvature reduce when the first-layer compaction force is increased. Their study also showed that the layer sequence has a significant effect on the tensile strength of the tablets, such that the tablets with brittle material as the first layer attain higher interfacial strength. However, the presence of different materials between the adjacent layers can also cause fracture and split at the interface due to inhomogeneous residual stress distribution in the layers [INM 07, INM 09]. Using a three-point bending experiment (Figure 5.1(b)), Podczeck and Al-Muti [POD 10, POD 06] showed that the modulus of elasticity and the size of particles influenced the overall strength of layered tablets/beams. They found that if the material forming the lower layer was more elastic, then the beam strength was reduced due to tension introduced into the system, acting especially at the layer interface and potentially causing partial or complete delamination. Larger differences in the particle size of the materials forming the tablet layers also resulted in an overall reduced tensile strength. In another flexure test experiment, Busignies *et al.* [BUS 13] showed that the fracture occurred at the interface or in the layers depending on the properties of the materials. In most cases, the highest tensile strength was obtained when the materials had similar elastic recovery. On the contrary, for materials with different elastic recoveries, the tensile strengths were reduced.

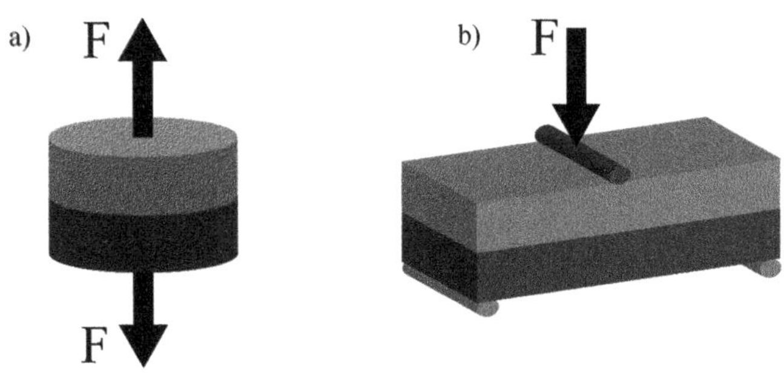

Figure 5.1. *Schematics of tensile strength tests for bilayer compacts: a) axial tensile strength test and b) three-point bending test*

Physical experiments can provide a lot of information about compaction of powders and the tensile strength of the compacts. However, in order to understand the evolution of the micromechanics, which occurs in the bulk of the powder bed, computational simulations are required. The experimental results discussed in the previous section can also be used to validate the results of the computational models.

5.2. Numerical simulation

Several computational methods have been developed to model compaction of monolayer tablets. Some of these methods are based on finite element method (FEM) [KLI 10], discrete element method (DEM) [SHE 04, MAR 04, OLS 13, CUN 79], or combined discrete-finite element method (DFEM) [ZHE 02, ZHA 11, KOY 11]. In FEM, usually only the global values of the physical properties are assumed, while in DEM the interaction between individual particles are accounted for, which enables modeling the heterogeneity of powders at the microscopic/particle scale. Hence, DEM gives more accurate particle scale details than continuum methods that consider global physical properties of the material. The discrete particle methods have been used to model compaction and tensile strength of monolayer tablets. In this section, first, we will show important details of the discrete particle model elaborated in Yohannes *et al.* [YOH 16], and then we will present how the monolayer powder compaction models can be improved to model bilayer powder compaction.

5.2.1. *Discrete particle simulations of powder compaction*

DEM was originally developed for applications pertaining to very small elastic deformations for rock mechanics application [CUN 79]. Since then, the DEM has seen several improvement and applications in a variety of engineering and scientific areas including modeling of plastically deformed particles [SHE 04, MAR 04, OLS 13] and bonding between particles [POT 04], which are relevant for powder compactions. As the main difference between monolayer and bilayer tablets is the presence of layer interface in the latter case, these computational methods can be used to model bilayer tablets if the contact between particles at the layer interface are accounted for.

During the compaction, unloading and ejection simulation, the contact force between the particles is computed based on the deformation and mechanical properties of the particles. In the simulations, the deformation of particle is computed from the overlap between contacting particles, and the mechanical properties include the plastic, elastic and bonding properties of the particles. A pair of contacting particles can be in a loading mode (when the particles are in contact for the first time), or unloading mode (when the contact force on a pair of contacting particles is decreasing) or reloading mode (when previously contacting particles are overlapping again) [YOH 16]. During loading mode, the particles follow a power-law plasticity model [MAR 04, MES 00, STO 97, STO 99],

$$\sigma = k\epsilon^{1/m}, \quad [5.1]$$

where k and m are particle plastic parameters, and ϵ and σ are the strain and stress of the particles, respectively. The contact force (P) between two particles is then computed as

$$P = k_p a^{2+1/m}, \quad [5.2]$$

where a is the radius of the contact area between the pair of contacting particles, $a = \sqrt{2c_2 \frac{\gamma}{\frac{1}{R_1}+\frac{1}{R_2}}}$. γ is the overlap between two contacting particles and R is the radius of the particles. The subscripts 1 and 2 refer to the pair of contacting particles. m is the average plastic parameter, $m = (m_1 + m_2)/2$. The parameter k_p in [5.2] is given as

$$k_p = K\pi\left(\frac{1}{k_1^m} + \frac{1}{k_2^m}\right)^{-\frac{1}{m}} \left(\frac{1}{R_1} + \frac{1}{R_2}\right)^{\frac{-1}{2m}-1} (2C_2)^{1+\frac{1}{2m}}, \quad [5.3]$$

where $K = 3 \times 6^{-1/m}$ and $c_2 = 1.43\exp\left(-\frac{0.97}{m}\right)$.

During the unloading of the contact force, P is computed as,

$$P = \frac{2k_p}{\pi} a_p^{2+1/m} \left[\arcsin\left(\frac{a}{a_p}\right) - \frac{a}{a_p}\sqrt{1 - \left(\frac{a}{a_p}\right)^2} \right] - \sqrt{8\pi\omega E_m a_p^{\frac{3}{2}}} \left(\frac{a}{a_p}\right)^{\frac{3}{2}}, \quad [5.4]$$

where a_p is the radius of the contact area at the maximum P during the loading phase, ω is the bonding energy per area and E_m is the effective Young's modulus.

During compaction of a powder bed, which consists of many particles, the type of each contact is evaluated and the appropriate equation is used in order to compute the forces on the particles. The computational process involves finding the optimum configuration (position of each particle), such that the potential energy of the system is kept the minimum. Yohannes *et al.* [YOH 16] have shown that this method can be used to model compaction of powders and diametral tensile strength tests for monolayer tablets. Their simulation results were validated using physical experiments.

5.2.2. *Discrete particle simulation of layered compacts*

As mentioned earlier, the main difference between monolayer compacts and layered compacts is the interface between layers for the layered compacts. To model the interaction at the interface, Yohannes *et al.* [YOH 17] suggested a slight modification for the interaction of particles at the interface. The modification is based on the fact that the particles at the surface of the first layer have already been deformed due to the first-layer compaction force. When the first-layer force is removed, the elastic deformation is recovered, but the plastic deformation remains. To account for this deformation during the interaction of particles across the interface, Yohannes *et al.* [YOH 17] assumed the following: 1) the radius of first-layer particles at the surface is smaller than the original particle radius by the amount of the plastic deformation (γ_p) and 2) the elastically recovered region, which is assumed as the “used part”, cannot support any bonding. Then, the overlap of particles across the interface (γ_{int}) is computed as

$$\gamma_{int} = (R_f + R_s - \gamma_p) - (X_f - X_s) \cdot \mathbf{n_{fs}}, \qquad [5.5]$$

where R_f is the original particle size of the particle in the first layer, R_s is the original particle size of the particle in the second layer, γ_p is the plastic deformation of the first-layer particle due to the first-layer force, X_f is the position vector of the center of the first-layer particle, X_s is the position vector of the second-layer particle, and n_{fs} is the unit normal vector between the centers of two particles.

Therefore, the contact force at the interface is developed only when $\gamma_{int} > 0$. For $0 < \gamma_{int} < \gamma_p$, it is assumed that the bonding between the particles is negligible and the contact force can only be compressive. If the contact is in a loading mode, [5.2] is used to compute the contact force, whereas, if the contact is in an unloading mode, the first term of [5.4] is used to compute the contact force. The bonding between the particles develops only when $\gamma_{int} \geq \gamma_p$. These assumptions about the interaction of particles, specifically the computation of the overlap for particles at the interface and the assumption about the bonding force between the particles, are the key factors that distinguish bilayer or multilayered tablets from monolayer tablets.

Simulations were run for lactose–lactose bilayer tablets based on these assumptions. Figure 5.2 shows the bilayer compaction process and the evolution of the bonding forces between contacting particles. Figure 5.2(a) shows the first layer being compacted and the bonding forces between particles of the first layer. The bonding between the particles still exists even after the first-layer compaction force is removed (Figure 5.2(b)). The bonding between the second-layer particles starts only after the second-layer force is applied (Figure 5.2(c)). Initially, the bonding force between the second-layer particles is much smaller than that between the first-layer particles. At the end of the compaction process, the bonding between particles of the first and second layers becomes almost equivalent (Figure 5.2(d)). These and other quantitative details are discussed in the next section.

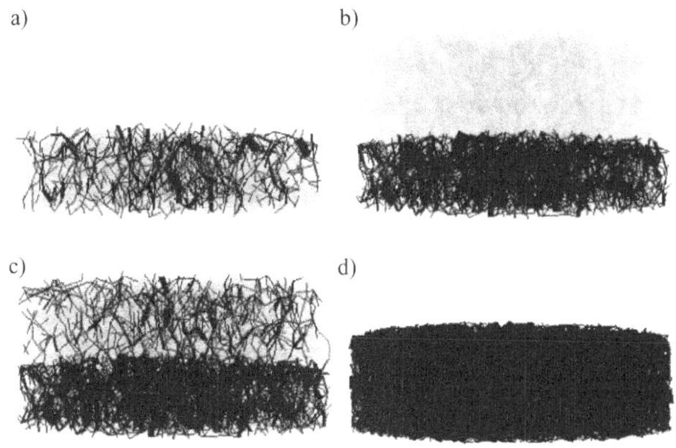

Figure 5.2. *The bonding force between particles: a) during compaction of the first layer; b) after the first layer is removed and second-layer particles are deposited on top of the first-layer particles; c) during the application of the second-layer force and d) at the maximum second-layer force. The thickness of the lines represents the magnitude of the bonding force between the particles*

5.3. Discussion

As mentioned in the previous section, one of the main advantages of the discrete particle method is that we can track the contact forces on each particle in detail. In this section, we present comparisons between contact forces and contact areas inside the two layers and at the interface of the layers. One way of checking the heterogeneity in the micromechanics of the tablet is to compute the probability distribution function (pdf) of the contact forces. Figure 5.3(b) shows the distribution of the contact forces inside the layers and at the interface of the layers for a lactose–lactose bilayer tablet after the tablet is ejected from the die. The contact forces at the layer interface are computed from all contacts between the first-layer and the second-layer particles. To have a reasonable comparison between the forces in the layers and at the interface, we only consider the contact forces across a given plane in the layers (Figure 5.3(a)). Though there is no external force applied on the tablet after ejection, there still exist some residual contact forces between the particles. From equilibrium criteria, we understand the summation of all these residual forces to be zero. In addition, we note that the distributions of the contact forces are similar for both layers and at the interface. The maximum compressive contact force is approximately 2 N and the maximum tensile force is about –1.2 N. The distribution resembles a Gaussian distribution with a mean around zero. The similarity in the contact forces indicates that there is no statistically noticeable difference between the contact forces.

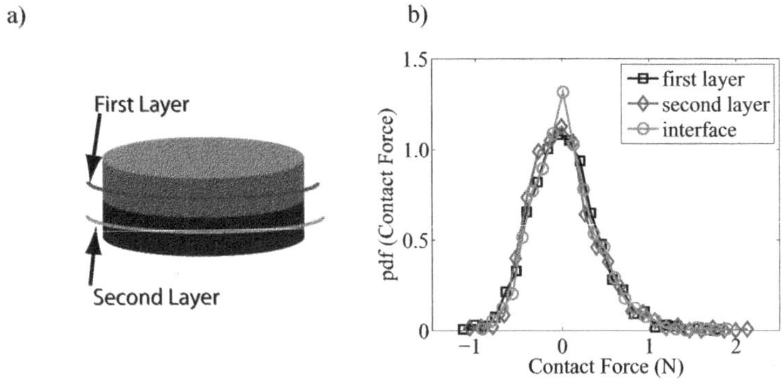

Figure 5.3. *a) Schematics that shows the planes at which the contact forces are computed and b) the distribution of contact forces in the first layer, second layer and at the interface for lactose–lactose bilayer after the tablet is ejected. Compressive forces are +ve and tensile forces are –ve. For a color version of the figure, see www.iste.co.uk/brancherie/microstructure.zip*

In addition to the pdf of the contact forces, we can also look into the orientation of the contact forces to investigate any preferential orientation of the contacts in the layers and at the interface. Figure 5.4 shows polar plots that represent the orientation and the magnitude of the contact forces. As shown in the schematics in Figure 5.4, the z-axis is along the axial direction, which is the direction of the applied compaction force. The horizontal plane is formed by the x–y axes. The orientation and the magnitude of the contact forces in the layers (only the result in the second layer is shown here) and at the layer interface are similar. The compressive forces (Figure 5.4(a) and (c)) have a maximum value of approximately 12 N mostly aligned between 30° and 60°, and symmetrical about z-axis and xy-plane. On the other hand, most of the tensile contact forces (Figure 5.4(b) and (d)) are aligned between 0° and 60°, which is also symmetrical about the z-axis and xy-plane for both layers and at the interface. These results are very similar to the orientation contact forces reported by Yohannes *et al.* [YOH 16] for monolayer compacts.

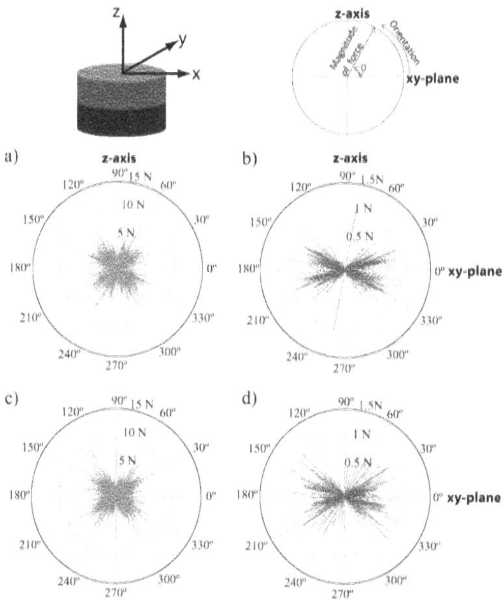

Figure 5.4. *The orientation of compressive and tensile contact forces in the second layer and at the interface during the maximum compaction force (relative density = 1.0): a) compressive contact force in the second layer, b) tensile contact force in the second layer, c) compressive contact force at the interface and d) tensile contact force at the interface. For a color version of the figure, see www.iste.co.uk/brancherie/microstructure.zip*

These contact forces do not represent the bonding between the particles. The bonding between the particles is equivalent to the force that is required to break the contact between the particles. The bonding force can be computed numerically from [5.4] or by taking the derivative of the equation with respect to the plastic contact radius a_p. Figure 5.5 shows the pdf of the bonding force between particles in the layers and at the interface. The bonding forces are computed in the regions as shown earlier in Figure 5.3(a). The bonding force ranges between 0 and –2.5 N. The distribution of the bonding force in the layers and at the interface is similar.

Based on the result presented in Figures 5.3–5.5, we can see that the distribution, orientation and the magnitude of the contact forces and bonding force are not sufficient to describe the difference in the tensile strength between the layers and at the interface. These quantities behave similarly at any point in the bilayer tablet. Perhaps, the appropriate measure of tensile strength is a cumulative measure of the bonding *forces*, which can be represented by the total bonding *area*. From our discrete particle simulations, the bonding area can be computed for all contacts between particles. Here, as described in section 5.2.2, we want to distinguish between the total contact area and the bonding area. The contact area is the total contact area between the particles; however, the existence of a contact area does not guarantee bonding between particles particularly for deformed particles of the first-layer particles at the layer interface.

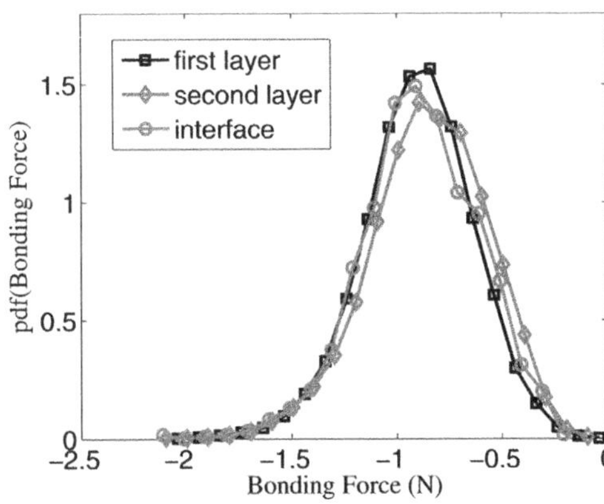

Figure 5.5. *The distribution of the bonding force in the first layer, second layer and at the layer interface for a bilayer tablet. For a color version of the figure, see www.iste.co.uk/brancherie/microstructure.zip*

In Figure 5.6, we show the contact area and bonding area for first-layer compaction pressures of 30 MPa, 60 MPa and 90 MPa. The contact area is normalized by the cross-sectional area of the smallest particle used in the simulation, which has a radius of 50 μm. In general, both the contact area and bonding area increase as the relative density of the tablets increases (relative density is computed as 1.0 – porosity). The increase in relative density corresponds to the increase in the second-layer pressure. For these tablets, a relative density of 1.0 corresponds to a second-layer pressure of approximately 250 MPa. The first layer has a higher contact and bonding area during the early stages of the application of the second-layer force. The area is higher because, as explained in section 5.2.2, the first layer has been already compacted during the application of the first-layer pressure. The contact area in the first layer increases by a factor of two when the first-layer compaction pressure is increased from 30 MPa to 90 MPa (compare Figure 5.6(a) and (e)). The initial contact area in the second layer and at the interface is zero since the second-layer pressure is not applied yet. The contact area in the second layer and at the interface increases continuously as the second-layer compaction pressure is applied, whereas the contact area in the first layer remains almost constant at the early stages of the application of the second-layer pressure. Once the contact areas in the second layer and at the interface approach the contact area in the first layer, the contact areas in the layers and at the interface increase at the same rate.

The bonding area is almost the same as the contact area for the first- and second-layer particles. However, the bonding area at the interface is less than the contact area, particularly when the first-layer pressure is 90 MPa. The lower bonding area indicates that the tensile strength at the interface is weaker at the interface. This result is similar to the experimental findings of Akseli *et al.* [AKS 13], in which they reported that, for higher first-layer compaction pressures, the bilayer tablets break at the interface during the axial tensile strength test. Actually, the tablets break at the interface with a pure delamination-type failure. For lower first-layer compaction pressures, Akseli *et al.* [AKS 13] found that the tablets could break in the first layer or second layer or at the interface, which, again, corresponds to our results where the bonding areas in the layers and at the interface are almost the same (e.g. Figure 5.6(b), when the first-layer pressure is 30 MPa). Had we considered only the contact areas, the simulations would not be able to predict the difference in the tensile strength of the layers and the interface.

Along the reduced area due to plastic deformation of the first-layer particles, the roughness of the first-layer surface plays an important role in the tensile strength of layered compacts [KOT 12]. Higher roughness provides more contact and bonding area at the interface. The roughness at the interface decreases as the magnitude of the first-layer compaction force is increased, since higher compaction forces cause significant particle rearrangement and deformation. Therefore, at higher first-layer compaction pressure, the contact area is further reduced by the smoothness of the top surface of the first layer. Actually, using the numerical model described here, Yohannes *et al.* [YOH 17] showed that the maximum bonding area is achieved at intermediate first-layer compaction pressure.

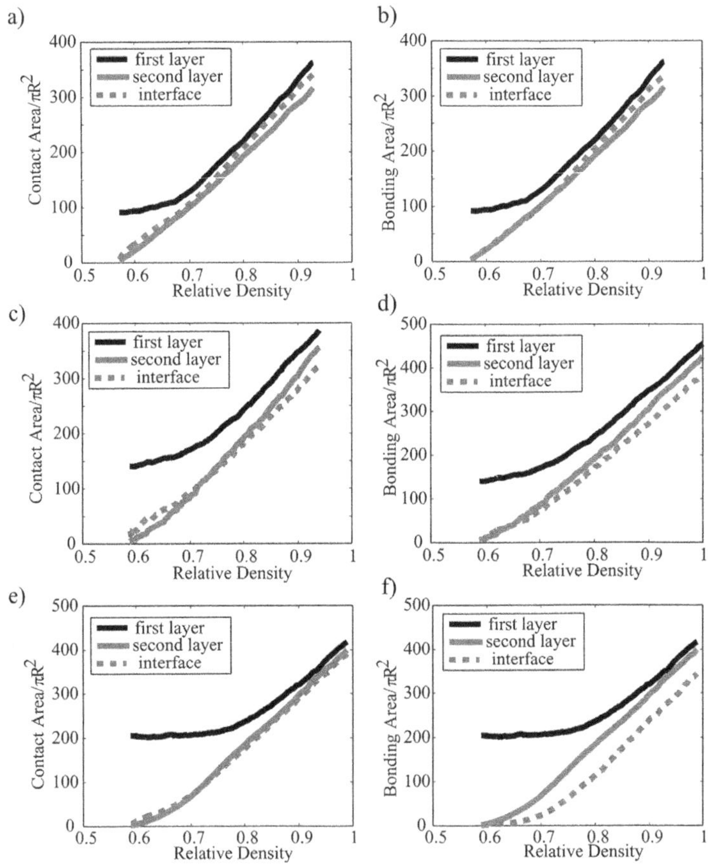

Figure 5.6. *Evolution of contact area and bonding area during the application of second-layer force. The tablets were compacted up to a relative density of 1.0, which corresponds to a second-layer pressure of 250 MPa. The first-layer pressure was 30 MPa for a) and b), 60 MPa for c) and d), and 90 MPa for e) and f). R is the radius of the smallest particle used in the simulations, which is 50 μm. For a color version of the figure, see www.iste.co.uk/brancherie/ microstructure.zip*

5.4. Conclusion

We presented a discrete particle-based modeling of bilayer powder compacts. The model is an extension of a model implemented earlier by Yohannes *et al.* [YOH 16] for monolayer tablets. The model for the bilayer powder compaction takes into consideration the plastic deformation of the first-layer particles, which were in direct contact with the punch during the application of the first-layer compaction pressure.

Using this model, we were able to show that the bonding area at the interface is significantly reduced for high first-layer compaction pressures. This reduced bonding area is the main cause of weak tensile strength at the interface, which was also observed in experiments. The reduced roughness at higher first-layer compaction pressures at the interface also plays an important role in reducing the tensile strength further. Despite the differences in the bonding area, other statistical parameters such as the distribution of the contact forces, the distribution of the bonding forces, and the orientation of the contact forces in the layers and at the interface are similar. Further, this model can be easily extended to multilayered powder compacts by repeating the model for deformation of the first-layer particles discussed in section 5.2.2 for all the layers.

5.5. Acknowledgements

We thank Admassu Abebe, Faranak Nikfar and Omar Sprockel for their fruitful discussions.

5.6. Bibliography

[ABD 04] ABDUL S., PODDAR S., "A flexible technology for modified release of drugs: multi layered tablets", *Journal of Controlled Release*, vol. 97, no. 3, pp. 93–405, 2004.

[ABE 14] ABEBE A., AKSELI I., SPROCKEL O. *et al.*, "Review of bilayer tablet technology", *International Journal of Pharmaceutics*, vol. 461, no. 12, pp. 549–558, 2014.

[ADO 97] ADOLFSSON A., Olsson H., Nystrom C., "Effect of particle size and compaction load on inter-particulate bonding structure for some pharmaceutical materials studied by compaction and strength characterization in butanol", *European Journal of Pharmaceutics and Biopharm*, vol. 44, pp. 243–251, 1997.

[AKS 13] AKSELI I., ABEBE A., SPROCKEL O. *et al.*, "Mechanistic characterization of bilayer tablet formulations", *Powder Technology*, vol. 236, pp. 30–36, 2013.

[ALM 08] ALMAYA A., ABURUB A., "Effect of particle size on compaction of materials with different deformation mechanisms with and without lubricants", *AAPS PharmSciTech*, vol. 9, no. 2, pp. 414–418, 2008.

[BUS 13] BUSIGNIES V., MAZEL V., DIARRA H. *et al.*, "Role of the elasticity of pharmaceutical materials on the interfacial mechanical strength of bilayer tablets", *International Journal of Pharmaceutics*, vol. 457, no. 1, pp. 260–267, 2013.

[CUN 79] CUNDALL P.A., STRACK O.D.L., "A discrete numerical model for granular assemblies", *Géotechnique*, vol. 29, pp. 47–65, 1979.

[DIV 11] DIVYA A., KAVITHA K., KUMAR M.R. *et al.*, "Bilayer tablet technology: an overview", *Journal of Applied Pharmaceutical Science*, vol. 1, pp. 43–47, 2011.

[EIC 09] EICHIE F.E., KUDEHINBU A.O., "Effect of particle size of granules on some mechanical properties of paracetamol tablets", *African Journal of Biotechnology*, vol. 8, no. 21, pp. 5913–5916, 2009.

[FIC 05] FICHTNER F., KE RASMUSON A., ALDERBORN G., "Particle size distribution and evolution in tablet structure during and after compaction", *International Journal of Pharmaceutics*, vol. 292, pp. 211–225, 2005.

[HEC 61] HECKLE R.W., "Density-pressure relationships in powder compaction", *Transactions of the Metallurgical Society of AIME*, vol. 221, pp. 671–675, 1961.

[HER 07] HERTING M.G., KLEINEBUDDE P., "Roll compaction/dry granulation: effect of raw material particle size on granule and tablet properties", *International Journal of Pharmaceutics*, vol. 338, pp. 110–118, 2007.

[INM 07] INMAN S., BRISCOE B., PITT K., "Topographic characterization of cellulose bilayered tablets interfaces", *Chemical Engineering Research and Design*, vol. 85, no. 7, pp. 1005–1012, 2007.

[INM 09] INMAN S., BRISCOE B., PITT K. *et al.*, "The non-uniformity of microcrystalline cellulose bilayer tablets", *Powder Technology*, vol. 188, no. 3, pp. 283–294, 2009.

[KHO 13] KHOMANE K.S., BANSAL A.K., "Effect of particle size on in-die and out-of-die compaction behavior of ranitidine hydrochloride polymorphs", *AAPS PharmSciTech*, vol. 14, no. 3, pp. 1169–1177, 2013.

[KLI 10] KLINZING G.R., ZavaLiangos A., CUNNINGHAM J. *et al.*, "Temperature and density evolution during compaction of a capsule shaped tablet", *Computers and Chemical Engineering*, vol. 34, no. 7, pp. 1082–1091, 2010.

[KOT 12] KOTTALA N., ABEBE A., SPROCKEL O. *et al.*, "Influence of compaction properties and interfacial topography on the performance of bilayer tablets", *International Journal of Pharmaceutics*, vol. 436, no. 12, pp. 171–178, 2012.

[KOY 11] KOYNOV A., AKSELI I., CUITIÑO A.M., "Modeling and simulation of compact strength due to particle bonding using hybrid discrete-continuum approach", *International Journal of Pharmaceutics*, vol. 418, pp. 273–285, 2011.

[MAR 04] MARTIN C., BOUVARD D., "Isostatic compaction of bimodal powder mixtures and composites", *International Journal of Mechanical Sciences*, vol. 46, no. 6, pp. 907–927, 2004.

[MCK 82] MCKENNA A., MCCAFFERTY D.F., "Effect of particle size on the compaction mechanism and tensile strength of tablets", *Journal of Pharmacy and Pharmacology*, vol. 34, no. 6, pp. 347–351, 1982.

[MES 00] MESAROVIC S.D., JOHNSON K., "Adhesive contact of elastic-plastic spheres", *Journal of the Mechanics and Physics of Solids*, vol.48, no. 10, pp. 2009–2033, 2000.

[OLS 13] OLSSON E., LARSSON P.-L., "A numerical analysis of cold powder compaction based on micromechanical experiments", *Powder Technology*, vol. 243, pp. 71–78, 2013.

[PAT 07] PATEL S., KAUSHAL A., BANSAL A., "Effect of particle size and compression force on compaction behavior and derived mathematical parameters of compressibility", *Pharmaceutical Research*, vol. 24, no. 1, pp. 111–124, 2007.

[POD 06] PODCZECK F., DRAKE K., NEWTON J. *et al.*, "The strength of bilayered tablets", *European Journal of Pharmaceutical Sciences*, vol. 29, pp. 361–366, 2006.

[POD 10] PODCZECK F., AL-MUTI E., "The tensile strength of bilayered tablets made from different grades of microcrystalline cellulose", *European Journal of Pharmaceutical Sciences*, vol. 41, no. 34, pp. 483–488, 2010.

[POT 04] POTYONDY D.O., CUNDALL P.A., "A bonded-particle model for rock", *International Journal of Rock Mechanics and Mining Sciences*, vol. 41, no. 8, pp. 1329–1364, 2004.

[SHE 04] SHENG Y., LAWRENCE C., BRISCOE B. *et al.*, "Numerical studies of uniaxial powder compaction process by 3d dem", *Engineering Computations*, vol. 21, pp. 304–317, 2004.

[STO 97] STORÅKERS B., BIWA S., LARSSON P.-L., "Similarity analysis of inelastic contact", *International Journal of Solids and Structures*, vol. 34, no. 24, pp. 3061–3083, 1997.

[STO 99] STORÅKERS B., FLECK N., MCMEEKING R., "The viscoplastic compaction of composite powders", *Journal of the Mechanics and Physics of Solids*, vol. 47, no. 4, pp. 785–815, 1999.

[SUN 06] SUN C.C., HIMMELSPACH M.W., "Reduced tabletability of roller compacted granules as a result of granule size enlargement", *Journal of Pharmaceutical Sciences*, vol. 95, no. 1, pp. 200–206, 2006.

[SUN 08] SUN C.C., "On the mechanism of reduced tabletability of granules prepared by roller compaction", *International Journal of Pharmaceutics*, vol. 347, pp. 171–172, 2008.

[YOH 15] YOHANNES B., GONZALEZ M., ABEBE A. *et al.*, "The role of fine particles on compaction and tensile strength of pharmaceutical powders", *Powder Technology*, vol. 274, pp. 372–378, 2015.

[YOH 16] YOHANNES B., GONZALEZ M., ABEBE A. *et al.*, "Evolution of the microstructure during the process of consolidation and bonding in soft granular solids", *International Journal of Pharmaceutics*, vol. 503, pp. 68–77, 2016.

[YOH 17] YOHANNES B., GONZALEZ M., ABEBE A. *et al.*, "Discrete particle modeling and micromechanical characterization of bilayer tablet compaction", *International Journal of Pharmaceutics*, vol. 529, pp. 597–607, 2017.

[ZHA 11] ZHANG J., ZAVALIANGOS A., "Discrete finite-element simulation of thermoelectric phenomena in spark plasma sintering", *Journal of Electronic Materials*, vol.40, no. 5, pp. 873–878, 2011.

[ZHE 02] ZHENG S., CUITIÑO A.M., "Consolidation behavior of inhomogeneous granular beds of ductile particles using a mixed discrete-continuum approach", *KONA Powder and Particle Journal*, vol. 20, pp. 168–177, 2002.

6

Microstructural Views of Stresses in Three-Phase Granular Materials

Granular materials are a specific class of materials used in various engineering fields – from materials engineering (metallic or ceramic powders) to civil engineering (granular soils), including food or pharmaceutical processes. Their specificity is directly related to their discrete microstructure which confers them very distinct features. Among other features, their granular skeleton reveals a pore network that may be filled with various fluids such as air, water, natural gas, oil, and so on. As with any other porous material, the possible presence of one or several fluids greatly affects the mechanical behavior of granular materials, which is evident with reference to the different properties of dry sand, fully water-saturated quicksand or wet sand. For instance, an apparent cohesion is observed in wet sand due to the mixture of air and water.

Due to a variety of internal forces, significant challenges readily arise in the mechanical modeling of three-phase (triphasic) granular materials such as wet sand. In the case of small pore sizes, the wetting fluid w (e.g. water in wet sand, or oil in oil-wet water-oil reservoirs) and the non-wetting fluid n (e.g. air in wet sand, or water in oil-wet water-oil reservoirs) typically show distinct fluid pressures u_w, u_n with $u_n > u_w$

Chapter written by Jérôme DURIEZ, Richard WAN and Félix DARVE.

because of the fluid-fluid surface tension γ_{nw}. Following Bishop's work [BIS 63], these fluid-related internal forces are classically lumped into an averaged fluid pressure, but such an approach may not capture all the intricacies of triphasic granular materials [BIS 63, NUT 08].

As an illustration, some experiments [SUN 04, SUN 07] on triphasic granular materials reveal surprising strains during a change in hydraulic conditions u_n, u_w. Upon wetting (decreasing $u_n - u_w$), partially water-saturated soils show some shear strains in addition to volumetric strains under specific stress conditions. Contrary to the volumetric ones, these shear strains may not be related to an isotropic stress loading such as a change in some averaged fluid pressure, revealing a possible flaw in this averaged pressure concept. Multiscale modeling approaches have indeed revealed a possible deviatoric component in the fluid contributions to the stress state of triphasic granular soils [HIC 07, SCH 09, WAN 14], which obviously presents a fundamental difference with averaged fluid pressures.

This deviatoric component is herein further investigated at the micro-scale from a combination of analytical homogenization and numerical simulations. An analytical homogenization approach [DUR 17a] is first introduced in section 6.1 to relate the stress state of triphasic granular materials to their microstructure. The obtained equations clearly show the possibility of a non-spherical stress contribution from the fluids, and this very deviatoric stress is measured from multiscale numerical experiments in section 6.3. To do so, a discrete element method (DEM) numerical model [DUR 17b], is presented in section 6.2, before being applied to access details of the microstructure. Finally, the classical question of effective stress in geotechnical engineering is addressed in section 6.4, demonstrating how the present microstructural approach enables us to relate the failure properties of triphasic granular materials to that of their dry counterparts.

6.1. Microstructural expression of triphasic total stresses

6.1.1. *Stress description within micro-scale volumes and interfaces of triphasic materials*

The micro-scale of triphasic granular materials first involves a solid phase s made of distinct particles p with the corresponding volume $V_s = \cup_p V_p$, as well as two immiscible fluid phases n, w showing the volumes V_n, V_w. The stress description of the fluid volumes obviously involves the stress tensors $\boldsymbol{\sigma_\alpha} = u_\alpha\, \boldsymbol{\delta}$ for $\alpha \in \{n, w\}$, with $\boldsymbol{\delta}$ the identity tensor and compressive stresses being herein considered as positive. As for the solid phase stress, it is classically computed as a volume average from tractions that apply along every particle surface S_p. Assuming quasi-static equilibrium without body forces, the following classical equality is actually used:

$$\frac{1}{V}\int_{V_s} \boldsymbol{\sigma_s}\, dV = \frac{1}{V}\sum_p \int_{V_p} \boldsymbol{\sigma_s}\, dV = \frac{1}{V}\sum_p \int_{S_p} (\boldsymbol{\sigma_s}\,\vec{n}) \otimes \vec{x}\, dS \qquad [6.1]$$

where $\vec{n}$ is the outward solid normal and position vectors $\vec{x}$ are expressed from adequate particle centroids. The solid phase stress is next obtained from equation [6.1] considering all tractions $\boldsymbol{\sigma_s}\,\vec{n}$ applied along S_p as solid contact forces, fluid pressures and surface tensions [DUR 16b, DUR 17a].

In addition to the bulk solid and fluid volumes with their classical stress tensors, it is important to distinguish the three interface surfaces $S_{\alpha\beta}$, $\alpha, \beta \in \{n, s, w\}$ and the associated surface tension fields within the material. In line with surface tension forces, these interfaces indeed carry specific energy, and thus they necessarily enter the mechanical description of triphasic materials. This necessity may be the most evident for the fluid-fluid interface S_{nw} and its surface tension γ_{nw}, but the solid-fluid surface tension $\gamma_{s\alpha}$ and interfaces $S_{s\alpha}$, $\alpha \in \{n, w\}$ are also *a priori* included, for the sake of generality. The description of surface tension phenomena requires an appropriate treatment precisely discussed, for example, by [GUR 75, CHA 95, DOR 06, GRA 07, DUR 16a, DUR 17a].

Tensorial surface stress fields $\boldsymbol{\pi}_{\alpha\beta}$ are first defined along each surface $S_{\alpha\beta}$ in order to account for the tensile membrane internal forces corresponding to surface tension:

$$\boldsymbol{\pi}_{\alpha\beta} = \gamma_{\alpha\beta} \left(\vec{n} \otimes \vec{n} - \boldsymbol{\delta} \right) \quad [6.2]$$

Equation [6.2] basically expresses the stress-like tensor $\boldsymbol{\pi}_{\alpha\beta}$ from the surface tension coefficient $\gamma_{\alpha\beta}$ and the projection tensor onto $S_{\alpha\beta}$, i.e. $(\boldsymbol{\delta} - \vec{n} \otimes \vec{n})$.

Second, surface tension also impacts the solid phase stress $\boldsymbol{\sigma_s}$. For example, along the solid-fluid surfaces $S_{s\alpha}$, $\alpha \in \{n, w\}$, an extra term proportional to $\gamma_{s\alpha}$ and the surface curvature of $S_{s\alpha}$ adds to the fluid pressure $u_\alpha \, \vec{n}$ in the boundary condition for the solid phase stress $\boldsymbol{\sigma_s} \, \vec{n}$ that appears in equation [6.1] [DOR 06, DUR 17a].

6.1.2. *Total stress derivation*

Accounting for the above, the total macroscopic stress $\boldsymbol{\Sigma}$ for the triphasic material is averaged from all the microscopic stress fields:

$$\boldsymbol{\Sigma} = \frac{1}{V} \left(\sum_{\alpha=n,s,w} \int_{V_\alpha} \boldsymbol{\sigma_\alpha} \, dV + \sum_{\alpha,\beta=n,s,w} \int_{S_{\alpha\beta}} \boldsymbol{\pi}_{\alpha\beta} \, dS \right) \quad [6.3]$$

Equation [6.3] eventually leads to the following expression, considering the non-wetting fluid pressure $u_n \boldsymbol{\delta}$ as a zero reference for the total stress [DUR 17a]:

$$\boldsymbol{\Sigma} = \frac{1}{V} \left[\sum_c \vec{f^c} \otimes \vec{l} - u_c \left(V_w \, \boldsymbol{\delta} + \int_{S_{sw}} \vec{n} \otimes \vec{x} \, dS \right) \right.$$

$$\left. - \gamma_{nw} \left(\int_{S_{nw}} (\boldsymbol{\delta} - \vec{n} \otimes \vec{n}) \, dS + \int_\Gamma \vec{\nu} \otimes \vec{x} \, dl \right) \right] \quad [6.4]$$

where $\vec{f^c}$ and $\vec{l}$ are the point contact forces and the branch vectors between contacting particles, $u_c = u_n - u_w$ is the capillary pressure corresponding to the geotechnical matric suction for granular materials and $\vec{\nu}$ is the inwards conormal to S_{nw} along the contact line Γ where the three phases meet (Figure 6.1).

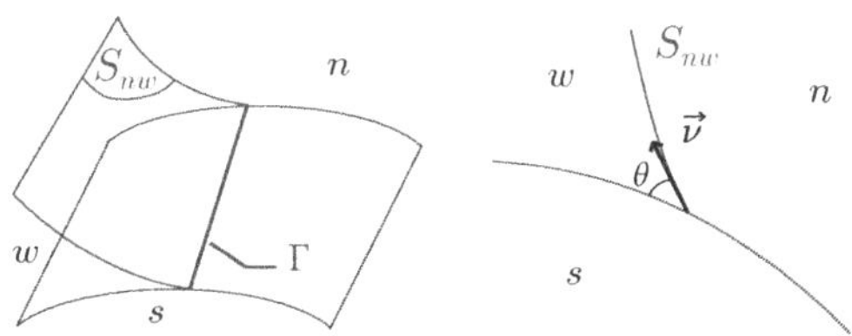

Figure 6.1. *Contact line Γ and inwards conormal $\vec{\nu}$ to S_{nw}*

The right-hand side terms of the total stress expression, equation [6.4], are further rearranged so as to distinguish a contact force stress contribution term denoted as the contact stress $\boldsymbol{\sigma}^{\boldsymbol{cont}}$:

$$\boldsymbol{\sigma}^{\boldsymbol{cont}} = \frac{1}{V}\sum_c \vec{f^c} \otimes \vec{l} \qquad [6.5]$$

from the fluid stress contribution term mentioned in the introduction, and from now on denoted as capillary stress $\boldsymbol{\sigma}^{\boldsymbol{cap}}$ [HIC 07, SCH 09, DUR 17a]:

$$\boldsymbol{\sigma}^{\boldsymbol{cap}} = -\frac{1}{V}\left[u_c\left(V_w\,\boldsymbol{\delta} + \int_{S_{sw}} \vec{n} \otimes \vec{x}\, dS\right)\right.$$

$$\left. + \gamma_{nw}\left(\int_{S_{nw}} (\boldsymbol{\delta} - \vec{n} \otimes \vec{n})\, dS + \int_{\Gamma} \vec{\nu} \otimes \vec{x}\, dl\right)\right] \qquad [6.6]$$

As such, the contact and capillary stresses add up to give the total stress

$$\boldsymbol{\Sigma} = \boldsymbol{\sigma}^{\boldsymbol{cont}} + \boldsymbol{\sigma}^{\boldsymbol{cap}} \qquad [6.7]$$

Several comments can be made on equations [6.4] and [6.6]. First, note that there is finally a zero stress contribution of the solid-fluid surface tension and interfaces in the considered case of rigid solid particles and vanishing solid-solid contact surfaces [NIK 13, DUR 17a].

Second, both the fluid pressure difference u_c and the fluid-fluid surface tension γ_{nw} enter the total and capillary stress expressions [6.4] and [6.6], which is at variance with the classical approaches such as Bishop's equation [BIS 63] that considers only fluid pressures.

The total and capillary stress expressions [6.4] and [6.6] are highly microstructure dependent. Defining the following microstructure tensors $\boldsymbol{\mu_{Vw}}, \boldsymbol{\mu_{Ssw}}, \boldsymbol{\mu_{Snw}}, \boldsymbol{\mu_{\Gamma}}$,

$$\boldsymbol{\mu_{Vw}} = V_w\, \boldsymbol{\delta} \qquad [6.8]$$

$$\boldsymbol{\mu_{Ssw}} = \int_{S_{sw}} \vec{n} \otimes \vec{x}\, dS \qquad [6.9]$$

$$\boldsymbol{\mu_{Snw}} = \int_{Snw} (\boldsymbol{\delta} - \vec{n} \otimes \vec{n})\; dS \qquad [6.10]$$

$$\boldsymbol{\mu_{\Gamma}} = \int_{\Gamma} \vec{\nu} \otimes \vec{x}\, dl \qquad [6.11]$$

capillary stress can be expressed as

$$\boldsymbol{\sigma^{cap}} = -\frac{1}{V}\Big[u_c\,(\boldsymbol{\mu_{Vw}} + \boldsymbol{\mu_{Ssw}}) + \gamma_{nw}\,(\boldsymbol{\mu_{Snw}} + \boldsymbol{\mu_{\Gamma}})\Big] \qquad [6.12]$$

Of these four microstructure tensors $\boldsymbol{\mu_X}$, it is obvious that only $\boldsymbol{\mu_{Vw}}$ is always spherical, leading to a possible deviatoric component of $\boldsymbol{\sigma^{cap}}$ in contradiction with the classical averaged fluid pressure expression. The deviatoric component may arise, e.g. from an anisotropic distribution of wetted surfaces S_{sw}, and is naturally expected to evolve during wetting because of configurational changes of these wetted surfaces S_{sw}. A wetting process under constant total stress, such as applied in the experiments by [SUN 04, SUN 07], may thus involve the deviatoric part of $\boldsymbol{\sigma^{cap}}$, and hence deviatoric contact stress $\sigma^{cont} = \Sigma - \sigma^{cap}$. Here, a possible scenario appears that would explain the above-mentioned deviatoric strains during wetting by associating wetting-induced changes to the deviatoric part of σ^{cont}. This is because the contact stress necessarily relates to the strains developed in the triphasic material, given that contact forces and particle relative displacements are related through an interparticle contact law.

The validity of such a scenario is further investigated in sections 6.3 and 6.4 where the anisotropy of capillary stress and the effective nature of the contact stress are actually measured. For this purpose, a numerical model describing altogether the strains, the microstructure and the stress of triphasic granular materials is necessary since experimental measurements of triphasic microstructures [BRU 13] are not yet precise enough to access the relevant microstructure tensors μ_X, and actually measure the capillary and contact stresses in the lab.

6.2. Numerical modeling of wet ideal granular materials

A DEM numerical model is appropriate for the task at hand since it allows us to directly probe into the detailed microstructure of granular materials. Also, the computational framework can be easily extended to triphasic conditions in a specific range of low wetting saturation [GIL 02, SCH 09]. Thus, a DEM model proposed in [DUR 17b] is herein adopted, in which a precise description of interparticle capillary bridges can be afforded so as to compute the macroscopic mechanical behavior of triphasic granular materials. The model applies to ideal granular materials with spherical particles at low wetting saturation values within the pendular regime where the wetting fluid is distributed into distinct capillary bridges, or so-called menisci, that bond distinct particle pairs.

6.2.1. *DEM description of fluid microstructure*

According to a computational paradigm proposed in [SCH 09], pendular regime capillary bridges (Figure 6.2) are calculated in the

DEM model from the Laplace-Young equation that defines the bridge geometry by relating the capillary pressure u_c and the surface tension γ_{nw} to the curvature $\mathrm{div}\vec{n}$ of the bridge profile, i.e. of the meniscus fluid-fluid interface:

$$u_c = \gamma_{nw}\,\mathrm{div}\vec{n} \qquad [6.13]$$

where $\vec{n}$ is the normal to the profile pointing to the wetting fluid.

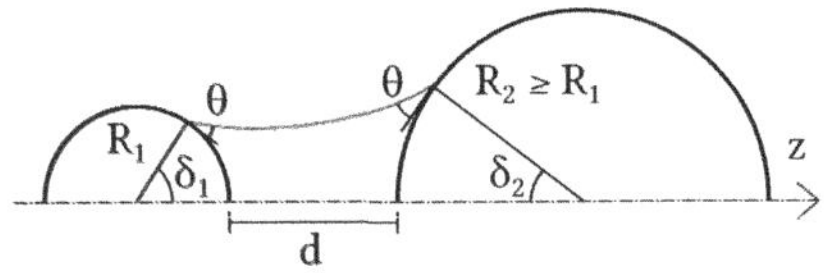

Figure 6.2. *Pendular regime capillary bridge*

In the present case of axisymmetric configurations (Figure 6.2), an adequate numerical procedure [LIA 93] allows the solution of equation [6.13] to fully characterize the meniscus geometry for an imposed capillary pressure, considering given particle radii R_1, R_2, interparticle distance d and contact angle θ (Figure 6.2). An extension of this procedure in [DUR 17b] enables us to compute a salient meniscus feature such as the microstructure tensor $\boldsymbol{\mu}_{\boldsymbol{Snw}}$ related to the fluid-fluid interface of each meniscus. Note that the spherical part of this microstructure tensor as m_{Snw} is directly proportional to the interface surface since

$$m_{Snw} = \frac{1}{3}\mathrm{tr}(\boldsymbol{\mu}_{\boldsymbol{Snw}})$$

$$= \frac{1}{3}\int_{Snw} \mathrm{tr}\,(\boldsymbol{\delta} - \vec{n}\otimes\vec{n})\;dS = \frac{1}{3}\int_{Snw} 2\,dS = \frac{2}{3}\,S_{nw} \qquad [6.14]$$

Also, a non-zero deviatoric part t_{Snw} usually emerges from $\boldsymbol{\mu}_{\boldsymbol{Snw}}$ in addition to its spherical part because of the meniscus orientation (Figure 6.3).

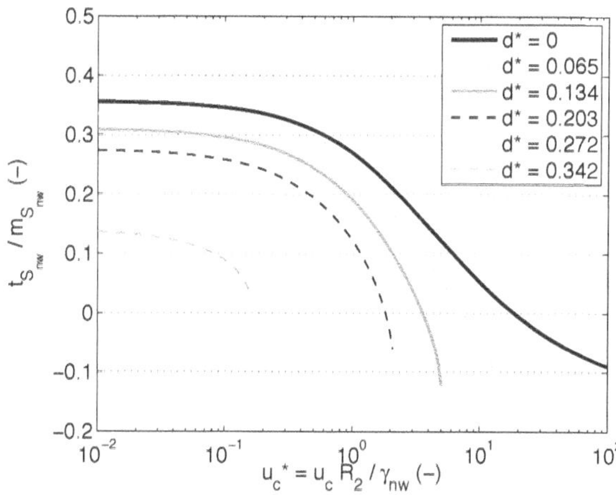

Figure 6.3. *Deviatoric nature of the fluid-fluid interface microstructure tensor μ_{Snw} at the capillary bridge scale. Bridges are considered in this figure to bond two particles of similar radii ($R_2/R_1 = 1.1$) with a contact angle of $10°$ and for different dimensionless capillary pressures u_c^* and interparticle distances $d^* = d/R_2$ [DUR 17a]. For a color version of the figure, see www.iste.co.uk/brancherie/microstructure.zip*

The deviatoric part t_{Snw} is defined as

$$t_{Snw} = \sqrt{\frac{3}{2}}\,||\boldsymbol{\mu}_{\boldsymbol{Snw}} - m_{Snw}\boldsymbol{\delta}|| \qquad [6.15]$$

and would be zero, e.g. in the case of cubic or spherical fluid-fluid interfaces.

As implemented in the YADE DEM code [SMI 15], this procedure eventually leads to a complete determination of the fluid distributions in granular materials. By construction, the computed distribution obeys uniform capillary pressure conditions i.e. thermodynamic equilibrium. As such, instead of computing transient pore flow, menisci are introduced in the DEM model by just solving the Laplace-Young equation [6.13] between all possible particle pairs for a given capillary pressure value. Uniform capillary pressure conditions may be difficult to ensure experimentally [GRA 15]; nevertheless, the results of the model still compare favorably with experimental measures of water retention behavior in granular materials (Figure 6.4). In Figure 6.4, S_r is the wetting saturation defined from the wetting fluid volume V_w and the void volume V_v as

$$S_r = \frac{V_w}{V_v} \quad [6.16]$$

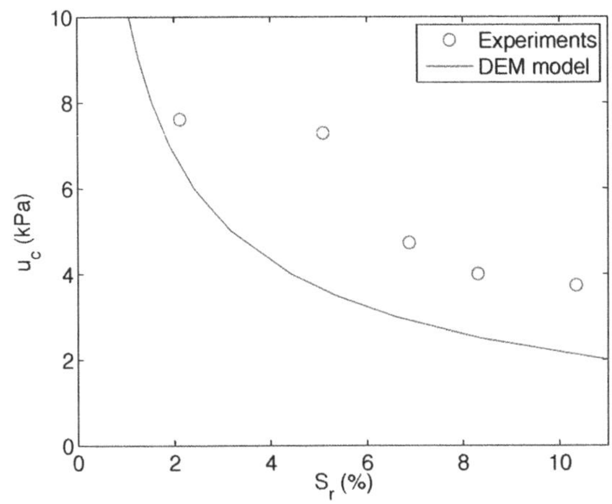

Figure 6.4. *Water retention behavior of a dense glass bead material ($n \approx 0.38; D \in [0.25; 0.3]$ mm). Experimental data are taken from [LIK 13]*

6.2.2. *DEM description of stress and strains*

Generally speaking, the DEM workflow introduces interaction forces between the solid particles and then applies the equations for rigid bodies movement to the particles, i.e. the discrete elements (DE) in a DEM model, to compute the particle relative displacements hence the macroscopic strains. Under triphasic conditions, those interaction forces shall be of two kinds, namely contact forces and capillary forces.

Contacting particles first interact naturally through contact forces $\vec{f^c}$ which are expressed from the particle relative displacements using classical elastic-plastic contact laws (for details, see [DUR 17b, WAN 15]).

Radial capillary attraction also appears in the model between capillary bonded particles, with the norm of the corresponding capillary force $\vec{f}^{cap}$ being readily expressed according to the known bridge's geometry:

$$\begin{aligned} \|\vec{f}^{cap}\| &= \pi\, R_1 \sin\delta_1\, (u_c\, R_1 \sin\delta_1 + 2\,\gamma_{nw} \sin(\theta+\delta_1)) \\ &= \pi\, R_2 \sin\delta_2\, (u_c\, R_2 \sin\delta_2 + 2\,\gamma_{nw} \sin(\theta+\delta_2)) \quad [6.17] \end{aligned}$$

where R_1, R_2 are the radii of the interacting particles 1 and 2, choosing $R_2 \geq R_1$ and δ_1, δ_2 are the filling angles on those particles 1, 2 (Figure 6.2). The bridges that carry capillary forces are created at each new contact, according to experimental observations [HER 05], with a configuration determined by solving the Laplace-Young equation [6.13]. Bridges and capillary forces subsequently stay in place even if particles separate, as long as the Laplace-Young equation has a solution. This is no longer the case beyond a limit distance which is a function of the imposed capillary pressure. Once this distance is reached, capillary bridges rupture and the capillary interaction is lost. Creations and ruptures of all bridges occur in the model under completely drained conditions, the water content and wetting saturation possibly evolving during suction-controlled simulations.

Note that the interaction forces may also serve to express the stress state of the DEM model, as demonstrated by Love [LOV 92] or the molecular dynamics virial theorem [PAR 54]. According to this paradigm, the total stress is directly explained by the interaction forces at contacts c and meniscus m, and the branch vectors $\vec{l}$ bridging interacting particles:

$$\mathbf{\Sigma} = \frac{1}{V}\left(\sum_c \vec{f}^c \otimes \vec{l} + \sum_m \vec{f}^{cap} \otimes \vec{l}\right) \quad [6.18]$$

The above equation [6.18] appears as an alternate measurement of the stress state, when compared with previous equations ([6.5], [6.6] and [6.7]). In particular, the capillary stress contribution is expressed in equation [6.18] from the resultant forces $\vec{f}^{cap}$ instead of distributed integrals in equation [6.6]. However, it was established in [DUR 16b, DUR 16a, DUR 17a] that these two ways of measuring capillary stresses are fully equivalent in the considered pendular regime. This astonishing equivalence stems from the following equality that applies to any capillary bridge and was demonstrated in [DUR 17a] from the Laplace-Young equation and meniscus mechanical equilibrium:

$$u_c \left(3\,V_m + \sum_{i=1,2} R_i\, S_{i,sw} \right) + \gamma_{nw} \left(2\,S_{nw} + \sum_{i=1,2} R_i\,\Gamma_i\, \sin\theta \right)$$
$$= \left(u_c\, \pi R_1{}^2 \sin^2 \delta_1 + \gamma_{nw}\, \Gamma_1\, \sin(\theta + \delta_1) \right) (R_1 + R_2 + d) \qquad [6.19]$$

Equation [6.19] links together capillary pressure, surface tension and meniscus geometric quantities such as the meniscus volume V_m, the wetted surfaces $S_{i,sw}$ and the contact lines lengths Γ_i along the two particles $i = 1, 2$. As such, this equation could also be of beneficial use in the experimental field to validate independent micro-scale measurements of the menisci distribution.

6.3. Anisotropy of the capillary stress contribution

The deviatoric nature of capillary stress is now measured, thanks to the DEM model which is applied along different hydraulic and mechanical loading paths, under axisymmetric conditions. Because of the axisymmetric conditions, the various stress tensors $\boldsymbol{\Sigma}, \boldsymbol{\sigma^{cap}}, \boldsymbol{\sigma^{cont}}$ are described using the following classical p, q invariant quantities:

$$p = \frac{1}{3}\mathrm{tr}(\boldsymbol{\Sigma}) \qquad q = \sqrt{\frac{3}{2}}\,||\boldsymbol{\Sigma} - p\,\boldsymbol{\delta}|| \qquad [6.20]$$

$$p^{cap} = \frac{1}{3}\mathrm{tr}(\boldsymbol{\sigma^{cap}}) \qquad q^{cap} = -\sqrt{\frac{3}{2}}\,||\boldsymbol{\sigma^{cap}} - p^{cap}\,\boldsymbol{\delta}|| \qquad [6.21]$$

$$p^{cont} = \frac{1}{3}\mathrm{tr}(\boldsymbol{\sigma^{cont}}) \qquad q^{cont} = \sqrt{\frac{3}{2}}\,||\boldsymbol{\sigma^{cont}} - p^{cont}\,\boldsymbol{\delta}|| \qquad [6.22]$$

It is obvious that $p = p^{cap} + p^{cont}$ for what concerns the mean stresses, and a similar relation $q = q^{cap} + q^{cont}$ holds for the deviatoric components. The additivity of the deviatoric components to reach a net value follows from having the same principal axes for all stress tensors, in line with axisymmetric conditions, and the use of a negative sign for q^{cap}. Using such variables, the tensorial anisotropy of $\boldsymbol{\sigma}^{cap}$, for instance, is readily quantified in the q^{cap}/p^{cap} ratio.

6.3.1. *Mechanical loading*

Triaxial compressions ($\Sigma_I \geq \Sigma_{II} = \Sigma_{III} = 10$ kPa) are first considered under triphasic conditions imposed at capillary pressure values of 4 and 40 kPa. The numerical samples include spherical particles whose diameters range between 0.25 and 0.3 mm. As a result, the pore air-water mixture shows an initial wetting saturation close to 4% ($u_c = 4$ kPa) or 0.1% ($u_c = 40$ kPa), for a 10° contact angle. In its initial stage, the solid particle packing is isotropic with a porosity $n = V_v/V \approx 0.379$.

Then, anisotropy is subsequently induced during the triaxial loading with the contact directions orienting along the major principal direction. This solid phase anisotropy is classically described from the contact fabric tensor $\boldsymbol{F_c}$:

$$\boldsymbol{F_c} = \frac{1}{N_c} \sum_c \vec{n} \otimes \vec{n} \qquad [6.23]$$

where N_c is the number of interparticle contacts and $\vec{n}$ is the associated contact normal. The contact anisotropy is quantified from the deviatoric component t_{Fc} of $\boldsymbol{F_c}$ with t_{Fc} defined as

$$t_{Fc} = \sqrt{\frac{3}{2}}\, ||\boldsymbol{F_c} - \frac{1}{3}\mathrm{tr}(\boldsymbol{F_c})\,\boldsymbol{\delta}|| = \sqrt{\frac{3}{2}}\, ||\boldsymbol{F_c} - \frac{1}{3}\,\boldsymbol{\delta}|| \qquad [6.24]$$

The growing contact anisotropy directly reflects on the wetted surface microstructure because menisci locate mostly at contacts. This is especially true at higher capillary pressures when the rupture distance of bridges is infinitesimal. As such, the deviatoric nature of the microstructure tensors $\boldsymbol{F_c}$ and $\boldsymbol{\mu_{Ssw}}$ follows very close trends during the triaxial compression with 40 kPa capillary pressure (Figure 6.5).

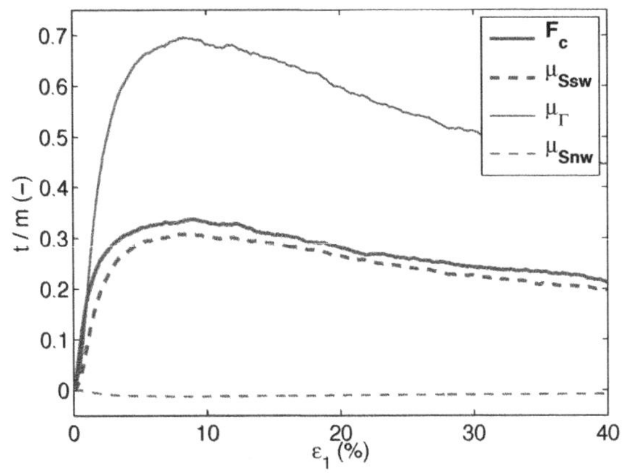

Figure 6.5. *Deviatoric nature of relevant microstructure tensors during a triphasic triaxial compression with* $u_c = 40$ *kPa*

An even greater deviatoric characteristic is measured for the contact line microstructure tensor $\boldsymbol{\mu_\Gamma}$, which relates to the used low contact angle value ($\theta = 10°$) – for a zero contact angle, $m_\Gamma = 1/3\,\mathrm{tr}(\boldsymbol{\mu_\Gamma})$ would be zero and the deviatoric nature t/m undefined. Finally, for what concerns the fluid-fluid interface, the corresponding microstructure tensor $\boldsymbol{\mu_{Snw}}$ is very close to being spherical at the sample scale even though it is not at the meniscus scale (Figure 6.3).

As expected, capillary stresses themselves also show a significant deviatoric nature because of microstructure anisotropy (Figure 6.6). This deviatoric nature is initially zero when the sample is isotropic, then grows to a peak occurring after the peak stress, and finally shows some reduction towards the critical state, in line with the evolutions of the solid phase fabric tensor $\boldsymbol{F_c}$ (Figure 6.5). Such results clearly demonstrate how capillary stress may deviate from an averaged fluid pressure, especially in the case of very low wetting saturation.

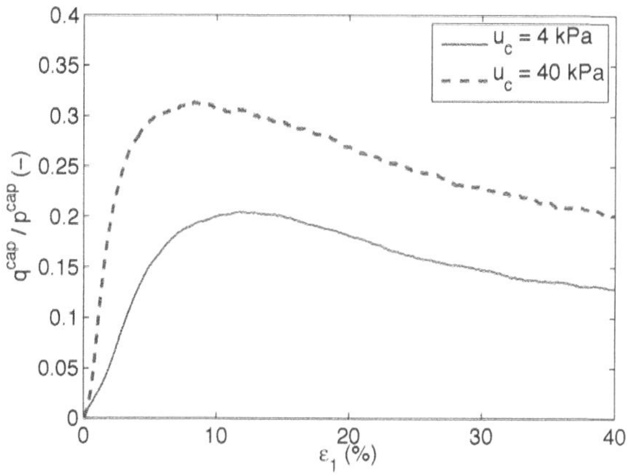

Figure 6.6. *Deviatoric nature of capillary stresses during two triaxial compressions*

6.3.2. *Hydraulic loading*

Turning now to hydraulic loading paths, note that the lab experiments of initially triphasic samples by [SUN 04, SUN 07] where wetting proceeds to full saturation cannot be reproduced by the model because of its restriction to the pendular regime. Instead, it is now considered a limited wetting of initially dry (air-saturated) samples. Such a wetting is applied to numerical samples with a mean particle diameter of 0.04 mm and a porosity close to 0.37 (for details, see [WAN 15]).

Two samples of the same material are considered, having sustained different loading histories in terms of different dry triaxial compressions with a 10 kPa confining pressure. They thus show different solid phase microstructures (Figure 6.7).

Sample 1 has been compressed to 0.1% axial strain only, i.e. a stress ratio $q/p \approx 0.85$. It shows an almost isotropic contact network. Sample 2, on the other hand, has been taken closer to peak stress, i.e. 0.3% axial strain and $q/p \approx 1.06$, and reveals a significant induced anisotropy. Triphasic conditions are subsequently imposed from these dry states, until a wetting saturation of 10% while keeping constant the total stresses p, q with corresponding non-zero q/p ratios. In this sense, the present procedure is somewhat similar to the experimental procedures followed in [SUN 04, SUN 07] where wetting is imposed under anisotropic total stresses and the actual soil samples certainly also reveal some induced anisotropy.

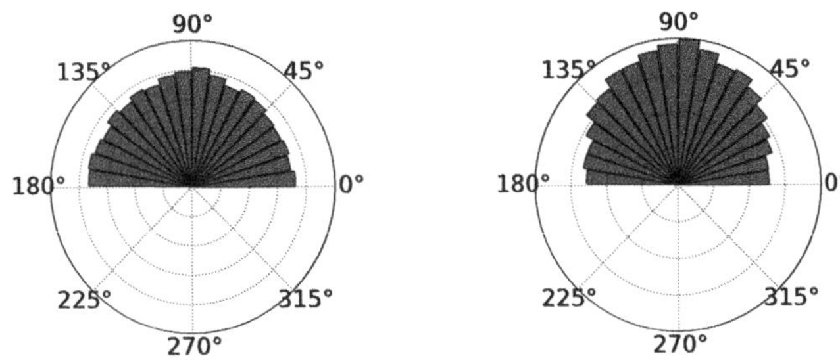

Figure 6.7. *Contact distribution of the considered samples 1 (left) and 2 (right)*

Because of the different solid microstructures, different capillary stresses develop in the triphasic material upon wetting (Figure 6.8). A deviatoric component q^{cap} of capillary stresses is especially significant in the anisotropic sample 2, whereas it is almost absent from the quasi-isotropic sample 1. It is also important to note that capillary stresses differ between samples 1 and 2 even though these samples show virtually identical soil water characteristic curves $u_c(S_r)$ [WAN 15].

As a result of the two different capillary stresses, wetting triggers distinct evolutions of the contact stress in samples 1 and 2. In particular, for sample 2, the deviatoric contact stress q^{cont} abruptly deviates from the total deviatoric stress at the dry-wet transition, a phenomenon that is barely visible in sample 1 (Figure 6.9).

As a possible result of the changes in the deviatoric contact stress, expected shear strains are actually measured transitioning from biphasic to triphasic conditions, with the greater shear strains for the more anisotropic sample 2 (Figure 6.10).

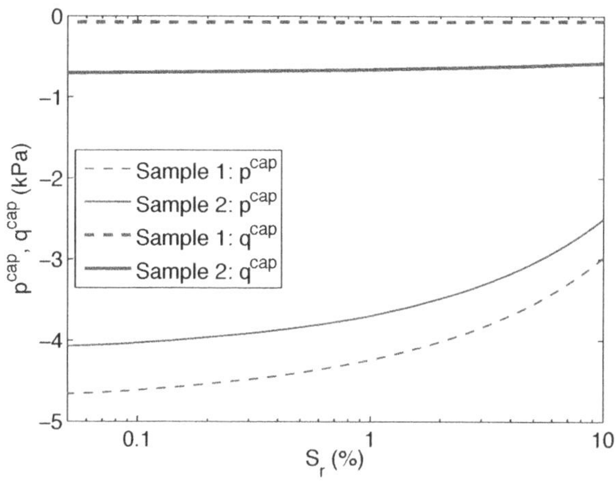

Figure 6.8. *Wetting-induced capillary stresses for samples 1 and 2*

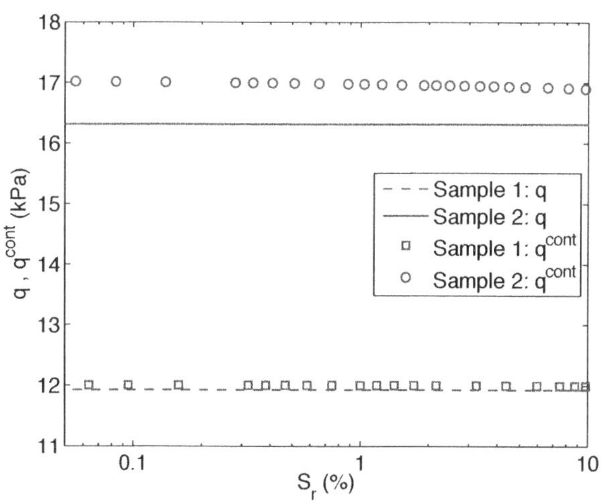

Figure 6.9. *Wetting-induced deviatoric contact stresses for samples 1 and 2. Prior to wetting both samples conformed* $q = q^{cont}$

The observed shear strains are of a limited magnitude that also applies to wetting-induced volumetric strains [WAN 15] for the granular material at hand. It is expected that other materials with a smaller particle size distribution would reveal greater wetting-induced strains.

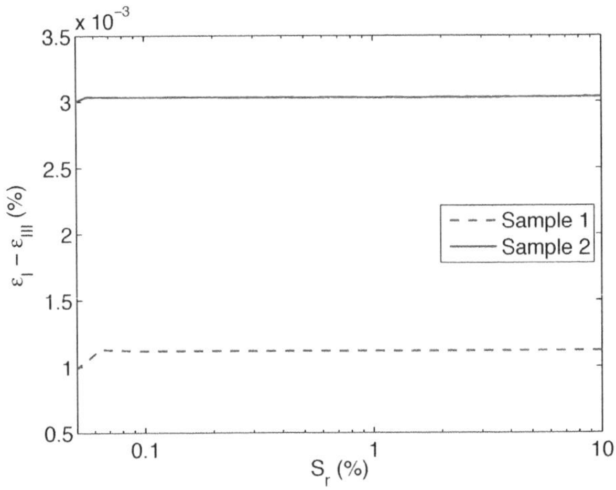

Figure 6.10. *Wetting-induced shear strains (with respect to the dry state) for samples 1 and 2*

6.4. Effective stress

A thorough explanation of wetting-induced strains would actually require the introduction of various constitutive relations, which may be expressed, among other choices, in an effective stress framework. The effective stress concept was initially developed in the biphasic case by Terzaghi and Biot [TER 36, BIO 41] who demonstrated that the same constitutive relations of air-saturated (dry) materials may apply to any biphasic (fluid saturated) condition, irrespective of the pore fluid pressure, provided an effective stress is considered instead of the total stress. In line with this approach, describing triphasic strains and failure through an effective stress variable and dry relations would be highly beneficial from a practical viewpoint. This in fact is the motivation for the capillary stress expressions proposed herein from micro-mechanics or earlier by Bishop [BIS 63], since the solid-phase related contact stress $\boldsymbol{\sigma}^{cont} = \Sigma - \boldsymbol{\sigma}^{cap}$ is expected to somehow govern the strains and failure of triphasic materials, as previously suggested. In fact, this is next rigorously tested for a material failure description.

The contact stress tensor is thus measured from the DEM model during triphasic conditions, from the knowledge of capillary stress and using the stress additivity equation [6.7]. While DEM also allows for a direct measurement of σ^{cont}, it is expected that experimental methods would more easily achieve such an indirect determination from the fluid phase microstructure (spatial arrangement), instead of attempting direct measurements of the contact forces. In order to assess the stress-strength effective nature of the contact stress, the limit states of the total and contact stress tensors during triphasic compressions (partial air-water saturations) are compared with the limit states observed in air-saturated conditions, for both dense and loose solid skeletons (Figure 6.11) [DUR 17c]. The triphasic conditions classically induce an apparent cohesion for the total stress limit states which contradicts the dry failure criterion. However, the contact stress limit states in biphasic or triphasic conditions all conform to the same cohesionless failure criterion for a given granular packing. This establishes the stress-strength "effective" nature of the contact stress to describe triphasic conditions from dry constitutive criteria, confirming previous results provided by, for example, [HIC 07, WAN 14].

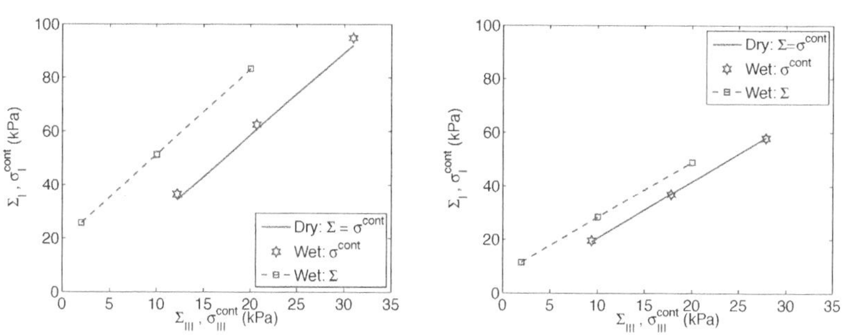

Figure 6.11. *Stress-strength effective nature of the contact stress tensor for dense (left) or loose (right) granular materials [DUR 17c]*

6.5. Conclusion

This chapter aimed at establishing the main reasons why the classical expressions of Bishop's type are not appropriate for describing the capillary stress in triphasic granular media.

According to continuum theories, mixture theory can be applied only when interstitial fluids are perfectly miscible, thereby leading to an isotropic capillary stress represented by a spherical tensor. However, when the wetting interstitial fluid and the non-wetting one are separated by finite interfaces where surface stresses are necessarily present to ensure equilibrium, capillary stress becomes inextricably anisotropic. In this chapter, this anisotropy has been formally demonstrated by applying analytical treatments to homogenize a triphasic granular medium based on the Love-Weber relation that links the microscopic force field to the macroscopic total stress tensor. While the analytically derived total stress expression clearly indicates an anisotropic capillary stress, it is also consistent with both the limiting cases of dry and fully saturated conditions. For the sake of completeness, this anisotropy has been fully confirmed by using an independent methodology that is based on the DEM.

Moreover, the total stress expression appears as the sum of three major contribution terms. The first one is the solid phase stress due to the intergranular contact forces. It is usually called "effective stress" in geomechanics, and this denomination has been justified for what concerns failure description. The second term corresponds to the part of the total stress that is proportional to the capillary pressure and is endowed with a tensorial character, which is essential. A third term is eventually proportional to the surface tension coefficient and also anisotropic. As one of the new results, this third term is totally absent in the classical expressions of the total stress tensor for triphasic granular materials.

Lastly, as pointed above, two major aspects of triphasic granular media have been identified: (1) the anisotropy of capillary stress and (2) the presence of a membrane-related term due to surface tension. The clear understanding of these two aspects surely elucidates the questions surrounding Bishop's type or other widely accepted expressions of capillary stress that have been in use for nearly half a century, despite their repeated discrepancies with experiments.

6.6. Bibliography

[BIO 41] BIOT M.A., "General theory of three-dimensional consolidation", *Journal of Applied Physics*, vol. 12, no. 2, pp. 155–164, 1941.

[BIS 63] BISHOP A.W., BLIGHT G.E., "Some aspects of effective stress in saturated and partly saturated soils", *Géotechnique*, vol. 13, pp. 177–197, 1963.

[BRU 13] BRUCHON J.-F., PEREIRA J.-M., VANDAMME M. *et al.*, "X-ray microtomography characterisation of the changes in statistical homogeneity of an unsaturated sand during imbibition", *Géotechnique Letters*, vol. 3, no. 2, pp. 84–88, 2013.

[CHA 95] CHATEAU X., DORMIEUX L., "Homogenization of a non-saturated porous medium: Hill's lemma and applications", *Comptes rendus de l'Académie des sciences. Série II*, vol. 320, pp. 627–634, 1995.

[DOR 06] DORMIEUX L., KONDO D., ULM F.-J., *Microporomechanics*, John Wiley & Sons, 2006.

[DUR 16a] DURIEZ J., WAN R., "Stress in wet granular media with interfaces via homogenization and discrete element approaches", *Journal of Engineering Mechanics*, vol. 142, no. 12, pp. 04016099-1–04016099-9, 2016.

[DUR 16b] DURIEZ J., WAN R., "Subtleties in discrete element modelling of wet granular soils", *Géotechnique*, vol. 67, no. 4, pp. 365–370, 2016.

[DUR 17a] DURIEZ J., EGHBALIAN M., WAN R. *et al.*, "The micromechanical nature of stresses in triphasic granular media with interfaces", *Journal of the Mechanics and Physics of Solids*, vol. 99, pp. 495–511, 2017.

[DUR 17b] DURIEZ J., WAN R., "Contact angle mechanical influence for wet granular soils", *Acta Geotechnica*, vol. 12, no. 1, pp. 67–83, 2017.

[DUR 17c] DURIEZ J., WAN R., POURAGHA M., "Partially saturated granular materials: insights from micro-mechanical modelling", in VANDAMME M., DANGLA P., PEREIRA J.-M. *et al.* (eds), *Poromechanics VI: Proceedings of the Sixth Biot Conference on Poromechanics*, pp. 441–448, 2017.

[GIL 02] GILI J.A., ALONSO E.E., "Microstructural deformation mechanisms of unsaturated granular soils", *International Journal for Numerical and Analytical Methods in Geomechanics*, vol. 26, no. 5, pp. 433–468, 2002.

[GRA 07] GRAY W.G., SCHREFLER B.A., "Analysis of the solid phase stress tensor in multiphase porous media", *International Journal for Numerical and Analytical Methods in Geomechanics*, vol. 31, no. 4, pp. 541–581, 2007.

[GRA 15] GRAY W.G., DYE A.L., MCCLURE J.E. *et al.*, "On the dynamics and kinematics of two-fluid-phase flow in porous media", *Water Resources Research*, vol. 51, no. 7, pp. 5365–5381, 2015.

[GUR 75] GURTIN M.E., MURDOCH A.I., "A continuum theory of elastic material surfaces", *Archive for Rational Mechanics and Analysis*, vol. 57, no. 4, pp. 291–323, 1975.

[HER 05] HERMINGHAUS S., "Dynamics of wet granular matter", *Advances in Physics*, vol. 54, no. 3, pp. 221–261, 2005.

[HIC 07] HICHER P.-Y., CHANG C., "A microstructural elastoplastic model for unsaturated granular materials", *International Journal of Solids and Structures*, vol. 44, no. 7–8, pp. 2304–2323, 2007.

[LIA 93] LIAN G., THORNTON C., ADAMS M.J., "A theoretical study of the liquid bridge forces between two rigid spherical bodies", *Journal of Colloid and Interface Science*, vol. 161, no. 1, pp. 138–147, 1993.

[LIK 13] LIKOS W., JAAFAR R., "Pore-scale model for water retention and fluid partitioning of partially saturated granular soil", *Journal of Geotechnical and Geoenvironmental Engineering*, vol. 139, no. 5, pp. 724–737, 2013.

[LOV 92] LOVE A., *A Treatise on the Mathematical Theory of Elasticity*, Cambridge the University Press, Cambridge, 1892.

[NIK 13] NIKOOEE E., HABIBAGAHI G., HASSANIZADEH S. *et al.*, "Effective stress in unsaturated soils: a thermodynamic approach based on the interfacial energy and hydromechanical coupling", *Transport in Porous Media*, vol. 96, no. 2, pp. 369–396, 2013.

[NUT 08] NUTH M., LALOUI L., "Effective stress concept in unsaturated soils: clarification and validation of a unified framework", *International Journal for Numerical and Analytical Methods in Geomechanics*, vol. 32, no. 7, pp. 771–801, 2008.

[PAR 54] PARKER E.N., "Tensor virial equations", *Physical Review*, vol. 96, pp. 1686–1689, 1954.

[SCH 09] SCHOLTÈS L., HICHER P.-Y., NICOT F. *et al.*, "On the capillary stress tensor in wet granular materials", *International Journal for Numerical and Analytical Methods in Geomechanics*, vol. 33, no. 10, pp. 1289–1313, 2009.

[SMI 15] SMILAUER V. *et al.*, *Yade Documentation*, 2nd ed., The Yade Project, 2015.

[SUN 04] SUN D., MATSUOKA H., XU Y., "Collapse behavior of compacted clays in suction-controlled triaxial tests", *Geotechnical Testing Journal*, vol. 27, pp. 362–370, 2004.

[SUN 07] SUN D., SHENG D., XU Y., "Collapse behaviour of unsaturated compacted soil with different initial densities", *Canadian Geotechnical Journal*, vol. 44, no. 6, pp. 673–686, 2007.

[TER 36] TERZAGHI K., "The shearing resistance of saturated soils and the angle between the planes of shear", *Proceedings of the 1st International Conference on Soil Mechanics and Foundation Engineering*, Cambridge, vol. 1, pp. 54–56, 1936.

[WAN 14] WAN R., KHOSRAVANI S., POURAGHA M., "Micromechanical analysis of force transport in wet granular soils", *Vadose Zone Journal*, vol. 13, no. 5, pp. 1–12, 2014.

[WAN 15] WAN R., DURIEZ J., DARVE F., "A tensorial description of stresses in triphasic granular materials with interfaces", *Geomechanics for Energy and the Environment*, vol. 4, pp. 73–87, 2015.

7

Effect of the Third Invariant of the Stress Deviator on the Response of Porous Solids with Pressure-Insensitive Matrix

Most of the efforts in modeling ductile damage concern materials for which the matrix plastic behavior is governed by the von Mises yield criterion. However, for most isotropic materials, the plastic deformation depends on both invariants of the stress deviator. The objective of this work is to study how the deviations in plastic deformation from the von Mises criterion influence the rate at which porosity accumulates in a porous solid. To this end, we conduct a micromechanical analysis of model unit cells, considering that the matrix behavior is governed by the Drucker yield criterion [DRU 49]. This criterion is pressure insensitive, and involves both the second invariant and the third invariant of the stress deviator. The relative weight of these invariants on the behavior is described by a parameter c. If this parameter is zero, there is no effect of the third invariant, and the Drucker criterion reduces to the von Mises yield criterion. The study of the response of the porous materials is conducted for both tensile and compressive axisymmetric loadings. For a given value of the stress triaxiality, loadings characterized by two different values of the Lode parameter, μ_Σ (corresponding to

Chapter written by José Luis ALVES and Oana CAZACU.

either major stress axial ($\mu_\Sigma = +1$) or major stress lateral ($\mu_\Sigma = -1$)), are analyzed for materials with a matrix characterized by $c < 0$, $c = 0$ (von Mises material) and $c > 0$. Irrespective of the sign of the parameter c, it is shown that the yield locus of the porous material is centro-symmetric. As a consequence, the rate of void growth is faster for axisymmetric loading histories corresponding to $\mu_\Sigma = +1$ than for those corresponding to $\mu_\Sigma = -1$, while void collapse occurs faster for loadings at $\mu_\Sigma = -1$ than for those characterized by $\mu_\Sigma = +1$. However, the sign of c has a drastic influence on void evolution; the rate of void growth or collapse is much faster for materials characterized by $c \geq 0$ than for materials characterized by $c < 0$.

7.1. Introduction

The ability to accurately describe the evolution of voids is essential for predicting ductile failure [MCI 68]. Beginning with the pioneering studies of Needleman [NEE 72], Tvergaard [TVE 81], Koplik and Needleman [KOP 88], micromechanical finite-element (FE) analyses of unit cells with a matrix governed by the von Mises yield criterion have provided fundamental understanding of the mechanical response of porous media (e.g. Richelsen and Tvergaard [RIC 94], Zhang *et al.* [ZHA 01]; Srivastava and Needleman [SRI 12], Alves *et al.* [ALV 14], etc.). In particular, it was revealed that, even for axisymmetric loadings, the stress triaxiality alone is insufficient to characterize important growth and coalescence features. Very recently, Cazacu *et al.* [CAZ 13] demonstrated that the exact yield criterion of porous materials with the von Mises matrix and randomly distributed spherical voids ought to be centro-symmetric, and should involve a very specific coupling between the mean stress and the third invariant of the stress deviator. Using rigorous limit-analysis theorems, analytic yield criteria that capture the aforementioned features of the yielding behavior were developed (for axisymmetric loadings, see [CAZ 13]; for full 3-D loadings, see [REV 14]). As the yield criterion involves coupling between the mean stress and shear stresses, the rate of void growth and collapse is influenced by the third invariant of the stress deviator. Such combined effects of mean stress and third invariant on yielding and void evolution in a porous Mises material were confirmed by unit-cell model simulation results (e.g. [ALV 13, ALV 14]). Since, in all the above studies, the fully dense material is described by the von Mises yield criterion, the effects of the third invariant of the stress deviator on the dilatational response of the porous solid are due solely to the presence of voids.

The mechanical response of porous isotropic materials for which the matrix is incompressible, but displays tension–compression asymmetry in plastic deformation, was studied by Cazacu and Stewart [CAZ 09]. The matrix behavior was modeled with the isotropic form of Cazacu *et al.* yield criterion [CAZ 06], which accounts for the specificities of the plastic deformation induced by either deformation twinning or non-Schmid slip at the grain level. Specifically, this yield criterion is an odd function of the stress deviator and involves dependence on its two invariants. Although plastic flow in the matrix is incompressible (no dependence of the mean stress), it displays tension–compression asymmetry (i.e. difference in response between tensile and compressive loadings). The dependence of the criterion on the third invariant is due solely to this tension–compression asymmetry. It was shown that the effects of the tension–compression asymmetry of the matrix on the yielding of the porous polycrystal are very pronounced. Specifically, it was demonstrated that the yield surface of the porous solid does not display any type of symmetries (see also [LEB 12] for full-field calculations, and [ALV 15] for unit-cell calculations that confirm these unusual features of the dilatational response). Furthermore, it was demonstrated that there is a very strong coupling between the tension–compression asymmetry of the matrix and porosity evolution (e.g. [REV 13a] study on porosity evolution in notched specimens loaded in uniaxial tension using [CAZ 09] model, for verification of the model predictions through comparison with X-ray computed micro-tomography data, see [REV 13b]).

Tresca's yield criterion is a pressure-insensitive yield criterion that is represented by an even function in stresses that depends on both invariants of the stress deviator (for the expression of Tresca's yield criterion in terms of invariants, see, e.g. [LUB 08]). For a porous Tresca material, very recently, Cazacu *et al.* [CAZ 14a], Revil-Baudard and Cazacu [REV 14] have derived plastic potentials expressed in the stress and strain-rate space, respectively. It was shown that the shape of the octahedral plane cross-sections of the plastic potential of the porous solid evolves from a regular hexagon to a smooth triangle with rounded corners. This in turn affects every aspect of the mechanical response of the porous solid. Let us denote the stress triaxiality, $T_\Sigma = \Sigma_m / \Sigma_e$, where Σ_e is the von Mises effective stress and Σ_m is the mean stress. It was shown that the rate of void growth is much faster in a porous Tresca solid than in a porous Mises solid. For example, for axisymmetric loadings at a stress triaxiality $T_\Sigma = 2$, the difference is approximately 20%, see also [REV 14, CAZ 14b]. Only for purely deviatoric loadings (i.e. at zero mean stress) or purely hydrostatic loadings ($T_\Sigma = \pm\infty$), the response of a porous Mises material and that of a porous Tresca material is the same.

For fully dense metallic materials, a versatile isotropic yield criterion was proposed by Drucker [DRU 49]. Like the von Mises and the Tresca criterion, the Drucker yield criterion [DRU 49] is pressure insensitive and predicts the same response in tension and compression. Its expression is a smooth function of both invariants of the stress deviator, and involves a parameter c. Depending on the sign of the parameter c, the Drucker yield locus is either interior ($c > 0$) or exterior ($c < 0$) to the von Mises yield locus. If $c = 0$, the Drucker (1949) criterion reduces to the von Mises yield criterion.

In this work, we investigate the mechanical response of a porous solid with a matrix governed by the Drucker yield criterion [DRU 49]. Such a study is also of interest in view of engineering applications, given that, for most isotropic fully dense metallic materials, the Drucker yield criterion describes better the plastic response than either the von Mises or the Tresca criterion (e.g. [LOD 26, TAY 31, DRU 49]; for a historical survey on early experimental works on mild steel tubes subjected to complex loading paths such as torsion and bending, torsion–compression and torsion–tension tests, the readers are referred to [MIC 76]).

The objective of this work is to study how the deviations in plastic deformation from the von Mises criterion influence the rate at which porosity accumulates in a porous solid. Of particular interest is the role of the parameter c on the evolution of porosity, and how it affects the ductility of the porous material. The outline of the chapter is as follows. We begin with a brief presentation of the Drucker yield criterion [DRU 49]. Next, the micromechanical unit-cell model and the method of analysis are discussed. In section 7.3 simulation results for porous materials with an incompressible matrix characterized by different values of the parameter c are presented. For each porous material considered, the macroscopic loadings imposed are such that the principal values of the applied stress, $\Sigma_1, \Sigma_2, \Sigma_3$, follow a prescribed proportional loading history corresponding to a constant stress triaxiality T_Σ. Specifically, the response is investigated under axisymmetric loadings ($\Sigma_1 = \Sigma_2$), where the axial overall stress Σ_3 is adjusted so that a fixed value of the stress triaxiality is maintained. The numerical yield surfaces for the porous materials are generated. To investigate the influence of the loading history (in particular, the effects of the third invariant J_3^Σ of the applied stress) on void growth and void collapse for any given specified value of the stress triaxiality T_Σ, loadings corresponding to either major stress axial ($J_3^\Sigma > 0$ for the entire loading history) or to major stress lateral ($J_3^\Sigma < 0$ for the entire loading history) are considered. It is shown that the parameter c, which accounts for the effects of the third invariant of the stress deviator on the plastic deformation of the matrix, has a strong influence on all aspects of the dilatational response of the porous materials. The main findings are summarized in section 7.4.

7.2. Problem statement and method of analysis

7.2.1. *Drucker yield criterion for isotropic materials*

At moderate temperature, plastic deformation occurs mainly by glide of dislocations and possibly, depending on the material, by twinning. If the deformation by crystallographic slip obeys Schmid law (i.e. yielding of the single crystal depends only on the magnitude of the resolved shear stress), the yield stresses in tension and compression are equal. Hence, the yield locus in the octahedral plane (plane that passes through the origin and is perpendicular to the hydrostatic axis) must have six-fold symmetry. As already mentioned, the two most used isotropic yield criteria for isotropic materials deforming by slip obeying Schmid's law are the Tresca [TRE 64] and von Mises [MIS 13] criteria. According to Tresca's criterion, the intermediate principal stress has no effect on yielding, while the von Mises yield criterion gives equal weight to all principal stresses. Lode [LOD 26] reported extensive data designed to draw a comparison between the yield criteria of Tresca and von Mises. Results of the tests on several materials subjected to combined axial tension and internal pressure correlated better with the von Mises criterion, although systematic deviations were observed. These deviations were attributed to experimental errors and/or to the anisotropy of the materials tested. The question was settled in 1931 when Taylor and Quinney [TAY 31] demonstrated that the deviations from the von Mises criterion were indeed real and that, for most isotropic metallic materials, the data were located between the two criteria. These findings led to the conclusion that in order to better represent the plastic deformation of isotropic materials, the yield criterion must depend on both invariants of the stress deviator of the Cauchy stress tensor, $\boldsymbol{\sigma}' = \boldsymbol{\sigma} - \sigma_{\mathrm{m}}\mathbf{I}$; with $\sigma_m = tr(\boldsymbol{\sigma})/3$ (tr denotes the trace operator, i.e. $tr(\mathrm{A}) = \sum_{k=1}^{3} A_{kk}$). A yield criterion that involves both invariants of σ' was proposed by Drucker (1949). Its expression is:

$$\left(\sigma_1'^2 + \sigma_2'^2 + \sigma_3'^2\right)^3 / 8 - c\left(\sigma_1'\sigma_2'\sigma_3'\right)^2 = \tau_{\mathrm{Y}}^6 \qquad [7.1]$$

where $\sigma_1', \sigma_2', \sigma_3'$ are the principal values of σ', while τ_Y is the yield stress in pure shear.

The necessary and sufficient condition for the Drucker yield surface to be convex is $-27/8 \le c \le 2.25$ (see [CAZ 01]). A judicious choice of the parameter, c, allows a very good relative weighting of the influence of all principal stresses, $\sigma_1, \sigma_2, \sigma_3$, on yielding. Depending on the sign of c, the Drucker yield locus is either interior

(c > 0) or exterior (c < 0) to the von Mises yield locus ($c = 0$ in Equation [7.1]). As an example Figure 7.1 shows the $\sigma_3 = 0$ cross-sections of the Drucker yield surfaces for $c = -27/8$ and $c = +2.25$, in comparison with the von Mises one ($c = 0$). The stresses are normalized by Y_0, the yield stress in uniaxial tension. Note that for c different from zero, there is a strong dependence of the yield loci on the third invariant, as evidenced by their shape (strong departure of the shape of the yield loci from the von Mises ellipse which corresponds to $c = 0$).

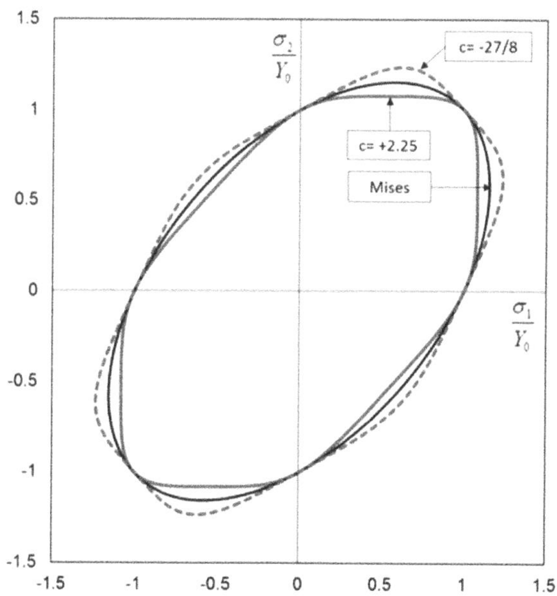

Figure 7.1. *Plane stress yield loci according to the isotropic Drucker yield criterion for: c = +2.25 and c = –27/8, in comparison with the Mises yield criterion (c = 0). σ_1, and σ_2 are the principal values of the Cauchy stress tensor*

7.2.2. *Unit cell model*

It is assumed that the porous isotropic material contains a regular and periodic 3-D array of initially spherical voids. The inter-void spacing is considered to be the same in any direction. The unit cell, which takes into account the periodicity of the porous medium, is initially cubic with side lengths 2 C_0 and contains a single spherical void of radius r_0 at its center. Thus, the initial porosity is:

$$f_0 = \frac{\pi}{6}\left(\frac{r_0}{C_0}\right)^3 \quad [7.2]$$

Cartesian tensor notations are used and the origin of the coordinate system is taken at the center of the void (see Figure 7.2(a)). Let u denote the incremental displacement between the current and reference configuration, and t the prescribed Cauchy stress vector, defined on the current configuration. Symmetry conditions are imposed on the planes x = 0, y = 0 and z = 0, respectively:

$$\begin{aligned} &u_1(0, y, z) = 0, \quad t_2(0, y, z) = 0, \quad t_3(0, y, z) = 0, \\ &u_2(x, 0, z) = 0, \quad t_1(x, 0, z) = 0, \quad t_3(x, 0, z) = 0, \\ &u_3(x, y, 0) = 0, \quad t_1(x, y, 0) = 0, \quad t_1(x, y, 0) = 0. \end{aligned} \quad [7.3]$$

Therefore, only one-eighth of the unit cell needs to be analyzed numerically (see Figure 7.2(b)). To simulate the constraints of the surrounding material, we enforce that the faces of the unit cell, which are initially planes parallel to the coordinate planes, remain planes and are shear free. The boundary conditions imposed on the faces of the unit cell are:

$$\begin{aligned} &u_1(C_0, y, z) = U_1^*(t), \quad t_2(C_0, y, z) = t_3(C_0, y, z) = 0 \\ &u_2(x, C_0, z) = U_2^*(t), \quad t_1(x, C_0, z) = t_3(x, C_0, z) = 0 \\ &u_3(x, y, C_0) = U_3^*(t), \quad t_1(x, y, C_0) = t_2(x, y, C_0) = 0. \end{aligned} \quad [7.4]$$

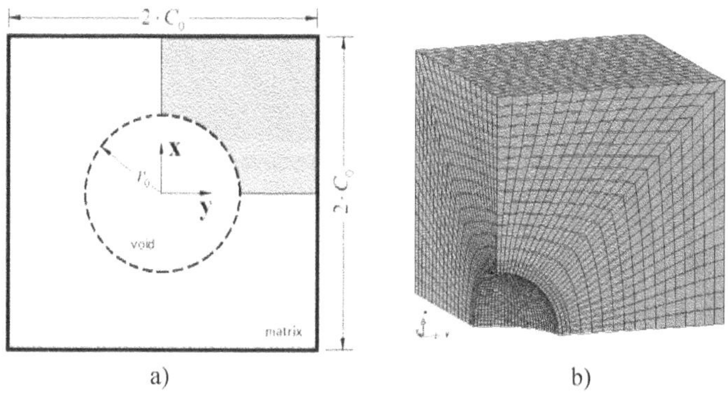

Figure 7.2. *a) Schematic two-dimensional projection of the three-dimensional cubic FE cell model adopted in this study; 2C0 and r0 denote the length of the undeformed cubic cell and the initial radius of the spherical void, respectively. b) Finite-element mesh of one-eighth of the unit cell with a spherical void at its center (mesh composed of 12150 8-node hexahedral elements)*

The void is considered to be traction-free. The macroscopic true stresses Σ_1, Σ_2,Σ_3 are defined as:

$$\Sigma_1 = \frac{1}{C_2 C_3}\int_0^{C_2}\int_0^{C_3} t_1 \, dzdy$$

$$\Sigma_3 = \frac{1}{C_1 C_2}\int_0^{C_1}\int_0^{C_2} t_3 \, dxdy \qquad [7.5]$$

$$\Sigma_2 = \frac{1}{C_1 C_3}\int_0^{C_3}\int_0^{C_1} t_2 \, dzdx$$

where $C_i = C_0 + U_i^*$ are the current cell dimensions. The porous material being isotropic, its mechanical response is fully characterized by the isotropic invariants of the overall stress, i.e.

$$\Sigma_m = \frac{1}{3}(\Sigma_1 + \Sigma_2 + \Sigma_3); \quad \Sigma_e = \sqrt{\frac{3}{2}\left(\Sigma_1'^2 + \Sigma_2'^2 + \Sigma_3'^2\right)}; \quad J_3^\Sigma = \Sigma_1'\Sigma_2'\Sigma_3' \qquad [7.6]$$

where $\Sigma_i' = \Sigma_i - \Sigma_m$, i = 1...3. In the analysis, we will refer to the following combinations of the stress invariants: the von Mises stress triaxiality ratio, T_Σ, and the Lode parameter μ_Σ, defined in terms of stress invariants as

$$\mu_\Sigma = \frac{27}{2} \cdot \frac{J_3^\Sigma}{(\Sigma_e)^3} \qquad [7.7]$$

The time histories of the displacements, $U_1^*(t)$, $U_2^*(t)$, and $U_3^*(t)$ in equation [7.4] are determined by the analysis in such a way that the macroscopic Cauchy stresses Σ_1, Σ_2, Σ_3 follow a prescribed proportional loading given by:

$$\frac{\Sigma_1}{\Sigma_2} = 1 \text{ (axisymmetric loadings) and } \frac{\Sigma_1}{\Sigma_3} = \rho \qquad [7.8]$$

where ρ is a prescribed constant such as to ensure that the stress triaxiality ratio, T_Σ, has a fixed specified value for the entire deformation process. The overall (macroscopic) principal strains and the macroscopic von Mises equivalent strain E_e are calculated as follows:

$$E_1 = \ln\left(\frac{C_1}{C_0}\right), \; E_2 = \ln\left(\frac{C_2}{C_0}\right), \; E_3 = \ln\left(\frac{C_3}{C_0}\right)$$

$$E_e^2 = \frac{2}{3}\left(E_1^2 + E_2^2 + E_3^2\right) \qquad [7.9]$$

where C_0 and C_i, $i = 1...3$ are the initial and current cell dimensions.

In the finite-element implementation of the above boundary value problem, all degrees of freedom with the same prescribed displacement of the type U_i^* belonging to the same planar bounding surface of the cubic surface are associated in the global stiffness matrix, and the equations of all these degrees of freedom are replaced by only one unknown variable. In this way, it is ensured that all initially planar boundary surfaces remain strictly flat during the entire loading history. Additionally, for each time increment and for all equilibrium cycles, the three imposed macroscopic forces on each planar bounding surface of the cubic cell are continuously updated in order to ensure the specified macroscopic Cauchy stress ratios on the final equilibrated configuration; the macroscopic non-equilibrated forces are introduced in the fully implicit Newton-Raphson algorithm in order to improve its convergence rate. Finally, a convergence criterion imposes that, for each planar surface, the ratio between the norm of the difference between the prescribed and effective macroscopic forces and the norm of the prescribed macroscopic force must be smaller than 0.001.

As mentioned, we focus on the study of the mechanical response of the porous solids for axisymmetric tensile loadings. At the end of each time increment, the condition of constant proportionality between the true stresses (equation [7.8]) is rigorously verified, so it is ensured that the macroscopic stress triaxiality, T_Σ, remains constant throughout the given deformation history. For each specified value of the stress triaxiality T_Σ, calculations are performed for the two possible ordering of the principal values, i.e. either: two principal values of the stress deviator $\boldsymbol{\Sigma}'$ are *compressive*: $\Sigma_1' = \Sigma_2' \leq 0$, but the maximum principal value $\Sigma_3' \geq 0$ is *tensile* (i.e. $\mu_\Sigma = 1$ and $J_3^\Sigma > 0$ for the entire deformation history), or two principal values of $\boldsymbol{\Sigma}'$ are *tensile*: $\Sigma_1' = \Sigma_2' \geq 0$, but *the minor principal value* $\Sigma_3' \leq 0$, which is *compressive (negative),* has the largest absolute value (i.e. $\mu_\Sigma = -1$ and $J_3^\Sigma < 0$ for the entire deformation history).

The void volume fraction, *f*, is evaluated at the end of each time increment as:

$$f = 1 - \frac{V_{matrix}}{V_{cell}}. \qquad [7.10]$$

In the above equation, $V_{cell} = C_1 C_2 C_3$, where C_i, i = 1…3 denote the current dimensions of the cell, while the volume of the deformed matrix, V_{matrix}, is determined directly from the finite element formulation, i.e.

$$V_{matrix} = \sum_{i=1}^{N_E} V_i \qquad [7.11]$$

where V_i is the current volume of the element i and N_E is the total number of finite elements in the mesh.

Using the FE unit cell model, the response of three porous materials with a matrix obeying the Drucker (1949) yield criterion (equation [7.1]) and the associated flow rule will be investigated. Specifically, three porous materials characterized by different values of the parameter c, but with the same initial void volume fraction, f_0 = 0.0104 (which corresponds to $r_0 / C_0 = 0.271$), will be considered. These are characterized by $c = -27/8$, $c = 0$ (von Mises) and $c = +2.25$.

In all the computations, all the other input material parameters are kept the same, i.e. the elastic properties and the material parameters involved in the isotropic hardening law describing the evolution of the matrix tensile yield strength with local equivalent plastic strain, $\bar{\varepsilon}^p$, i.e.

$$Y = A(\varepsilon_0 + \bar{\varepsilon}^p)^n \qquad [7.12]$$

where Y is the current matrix tensile flow stress, and A, n and ε_0 are material parameters.

The numerical values of the elastic parameters, the initial matrix yield stress, Y_0, and hardening exponent n, and ε_0 are respectively: $E/Y_0 = 500$, $\nu = 1/3$, $Y_0 = 400$ MPa, $A/Y_0 = 1.82$, $n = 0.1$. These numerical values for the material constants are typically used in FE unit cell calculations (e.g. [KOP 88]). It follows that all the differences in behavior between the porous materials are due solely to the effects of the third invariant on the plastic flow of the matrix, which are described by the parameter c. The FE analyses were performed with DD3IMP (Menezes and Teodosiu, 2000, Oliveira *et al.* 2008), an in-house quasi-static elastoplastic code with a fully implicit time integration scheme finite element solver.

Figure 7.2(b) shows the initial FE mesh of one-eighth of the unit cubic cell consisting of 12150 elements (8-node hexahedral finite elements; selective reduced integration technique, with 8 and 1 Gauss points for the deviatoric and volumetric parts of the velocity field gradient, respectively) and a total of 13699 nodes. A mesh refinement study was carried out to ensure that the results are mesh independent.

7.3. Results

7.3.1. *Yield surfaces and porosity evolution*

We begin by presenting the yield surfaces corresponding to porosity f = 1.04% of the porous Drucker materials with $c = 2.25$ and $c = -27/8$, respectively, in Figure 7.3 and Figure 7.5. The numerical yield points correspond to prescribed loads such that the stress triaxiality, T_Σ, is fixed. For either loading histories (i.e. either at $J_3^\Sigma > 0$ or $J_3^\Sigma < 0$), the stress triaxialities, T_Σ, range from 0 to $\pm\infty$. A total of 102 simulations were carried out for each material. Note that irrespective of the value of the parameter c, the yield surface of the porous Drucker solid is not symmetric with respect to the axis $\Sigma_m = 0$. Only for two loading conditions, namely purely deviatoric ($T_\Sigma = 0$) and hydrostatic loading ($T_\Sigma = \pm\infty$), the yield stress is insensitive to the third invariant of the applied load. For any given fixed value of the stress triaxiality T_Σ different from zero or infinity, there are two different yield points characterized by the same absolute value of J_3^Σ or μ_Σ (see equation [7.7] for the definition of μ_Σ).

At positive triaxialities (i.e. $\Sigma_m > 0$), for axisymmetric loadings at $J_3^\Sigma > 0$ ($\mu_\Sigma = 1$, major stress is axial), the yield stress is lower (see Figure 7.3) than in the case of loadings at $J_3^\Sigma < 0$ ($\mu_\Sigma = -1$, major stress is lateral). This in turn results in the rate of void growth being faster at $J_3^{\Sigma} > 0$ than at $J_3^{\Sigma} < 0$. As an example, see in Figure 7.4(a) the evolution of porosity for loadings at $T_{\Sigma} = 1$ for the porous material characterized by $c = 2.25$.

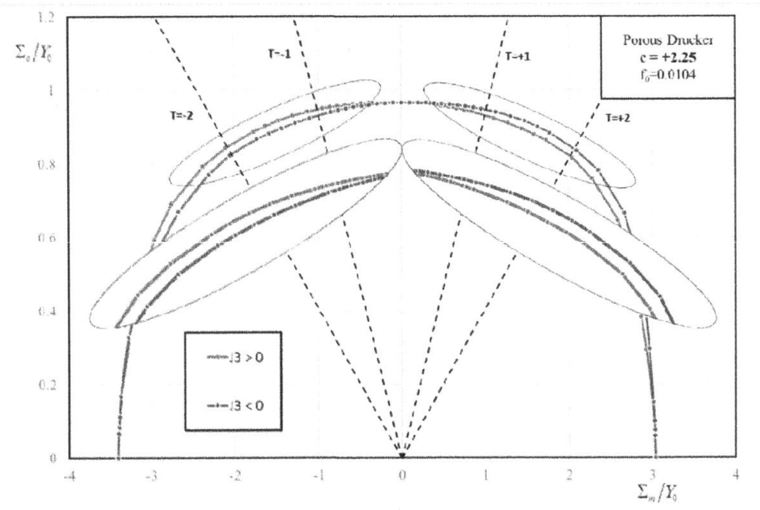

Figure 7.3. *The yield surface of a porous isotropic material with a matrix obeying the Drucker yield criterion with c=+2.25, determined by FE cell computations for macroscopic axisymmetric loadings corresponding to either J_3 <0 or J_3 >0. Initial void volume fraction f_0=0.0104*

On the other hand, for negative triaxialities (i.e. $\Sigma_m < 0$), the reverse holds true, i.e. for axisymmetric loadings at $J_3^{\Sigma} > 0$ the yield stress is larger than that for axisymmetric loadings at $J_3^{\Sigma} < 0$. Consequently, the rate of void collapse is faster for loadings at $J_3^{\Sigma} < 0$ than for loadings at $J_3^{\Sigma} > 0$. For example, Figure 7.4(b) shows, for the same material with $c = 2.25$, the void evolution for a triaxiality $T_{\Sigma} = -1$ and loadings at $J_3^{\Sigma} < 0$ and $J_3^{\Sigma} > 0$, respectively. The same conclusions can be drawn based on the numerical yield surface and the porosity evolution for the material characterized by $c = -27/8$ (see Figures 7.5–7.6).

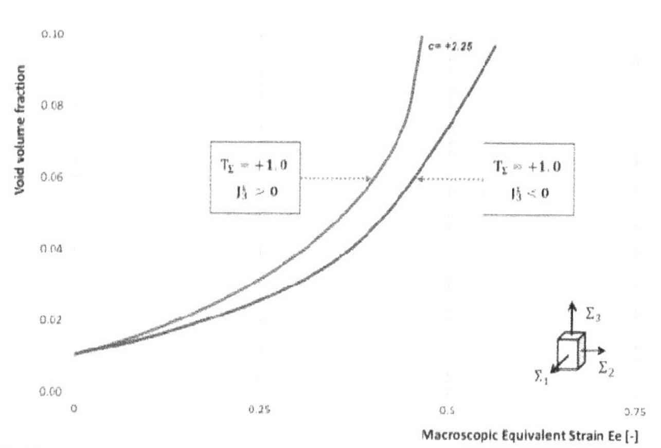

a)

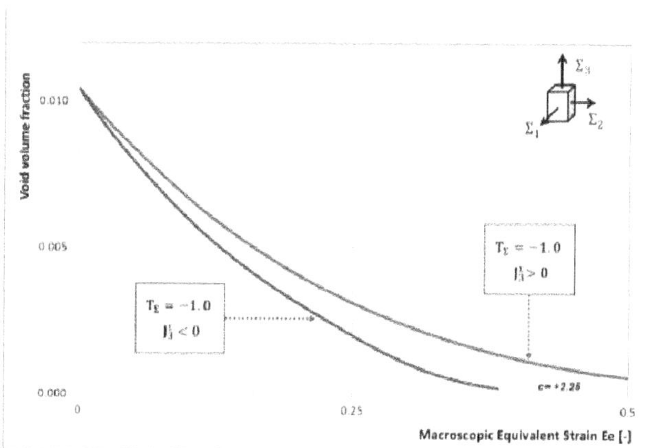

b)

Figure 7.4. *Evolution of the void volume fraction with the macroscopic von Mises equivalent strain Ee, obtained by FE cell calculations for an isotropic material with a matrix characterized by the Drucker yield criterion with parameter c = +2.25 for axisymmetric loadings corresponding to a fixed triaxiality of: (a) $T_\Sigma = +1$; and (b) $T_\Sigma = -1$. The rate of void growth is faster for loadings at $J_3 > 0$ than at $J_3 < 0$, while the rate of void collapse is faster for loadings at $J_3 < 0$ than for loadings at $J_3 > 0$*

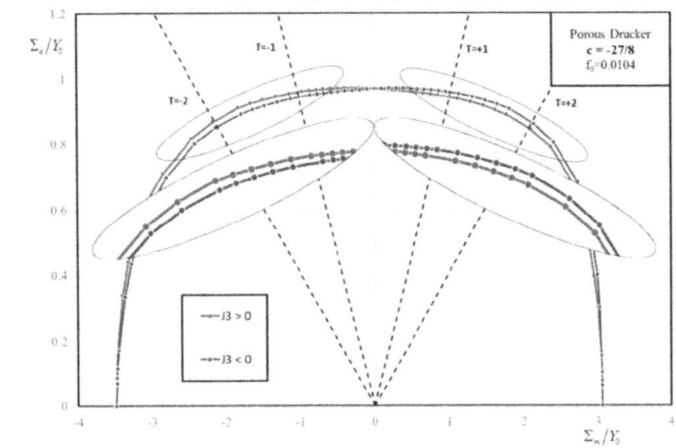

Figure 7.5. *The yield surface of a porous isotropic material with a matrix obeying the Drucker yield criterion with c = –27/8, determined by FE cell computations for macroscopic axisymmetric loadings corresponding to either $J_3 < 0$ or $J_3 > 0$. Initial void volume fraction $f_0 = 0.0104$*

The comparison between the porosity evolution of the porous Drucker materials for which the plastic deformation depends on J_3^Σ and that of a porous von Mises material ($c = 0$, plastic deformation in the matrix has no dependence on J_3^Σ) for the same loadings (at stress triaxialities $T_\Sigma = \pm 1$) is shown in Figures 7.7 and 7.8. The importance of the plastic deformation of the matrix is illustrated in these figures, in particular the very strong influence of the third invariant of the stress deviator described by the parameter c on porosity evolution. For $c = 2.25$, both the rate of void growth and the rate of void collapse are faster than those for a porous von Mises material. On the other hand, for a material with $c = -27/8$, the rate of void growth and that of void collapse are slower than those for a porous von Mises material (see Figure 7.9).

Let us discuss in detail the overall response of the three porous materials for loadings at $T_\Sigma = 1$ and $J_3^\Sigma > 0$. Figure 7.7 shows the void volume fraction vs. effective strain (Ee) curves, while Figure 7.9 depicts the rate of void growth. Note that void growth is fastest in the material with $c = +2.25$ and slowest in the material with $c = -27/8$. It is worth noting that, for the material with $c = -27/8$, the rate of void growth is lower and almost constant for most of the deformation processes. In contrast, for the porous von Mises material ($c = 0$) and for the porous material characterized by $c = 2.25$, damage accumulation is more gradual. To detect the onset of void coalescence (denoted by a black dot on the respective curves), we used the procedure outlined in [SRI 12], which is based on monitoring the evolution of the relative inter-void ligament size.

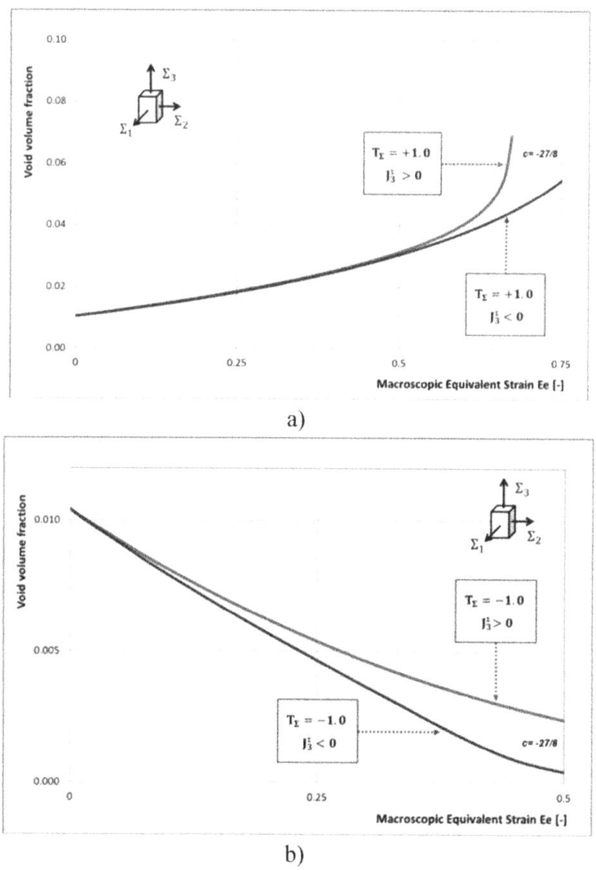

Figure 7.6. *Evolution of the void volume fraction with the macroscopic von Mises equivalent strain Ee, obtained by FE cell calculations for an isotropic material with matrix characterized by the Drucker yield criterion with parameter c = –27/8 for axisymmetric loadings corresponding to a fixed triaxiality of: (a)* $T_\Sigma = +1$ *and (b)* $T_\Sigma = -1$

To better understand the reasons for the very strong difference in the rate of void growth between the three porous materials, we further compare the local state fields corresponding to the same level of macroscopic true strain $E_e = 0.15$. Figure 7.10 shows the isocontours of the local equivalent plastic strain, $\bar{\varepsilon}^p$, corresponding to a macroscopic effective strain $E_e = 0.15$. The local equivalent plastic strain $\bar{\varepsilon}^p$ with the Drucker criterion given by equation [7.1] is calculated using the work-equivalence principle [HIL 87].

The white regions in Figure 7.11 represent the elastic zones, i.e. the non-plastified domains. It is worth noting that only in the material characterized by $c = -27/8$, almost the entire domain (cell) is plastified. In contrast, for the porous von Mises material ($c = 0$) and the porous material characterized by $c = +2.25$, there exists a zone in the vicinity of the void along the vertical axis of the cross-section (3-direction of axial stress (max stress), where yielding did not occur. Specifically, for the porous von Mises material ($c = 0$), the elastic zone is contiguous to the void while for the material characterized by $c = 2.25$, the elastic zone is slightly shifted upwards from the void. Further examination of the isocontours of the local equivalent plastic strain also shows very marked differences in terms of the heterogeneity of plastic deformation and the distribution of the plastic zones within the domain. Note that for the porous material with a matrix characterized by $c = -27/8$, the plastic deformation is more homogeneous than that in the other materials. To quantitatively estimate the heterogeneity in plastic deformation, we examine the ratio between the maximum local plastic strain in the entire domain, $\bar{\varepsilon}^p_{\max}$, and the average of the local plastic strain, $\left\langle \bar{\varepsilon}^p \right\rangle$ defined as:

$$\left\langle \bar{\varepsilon}^p \right\rangle = \frac{1}{V} \int_V \bar{\varepsilon}^p dV \qquad [7.13]$$

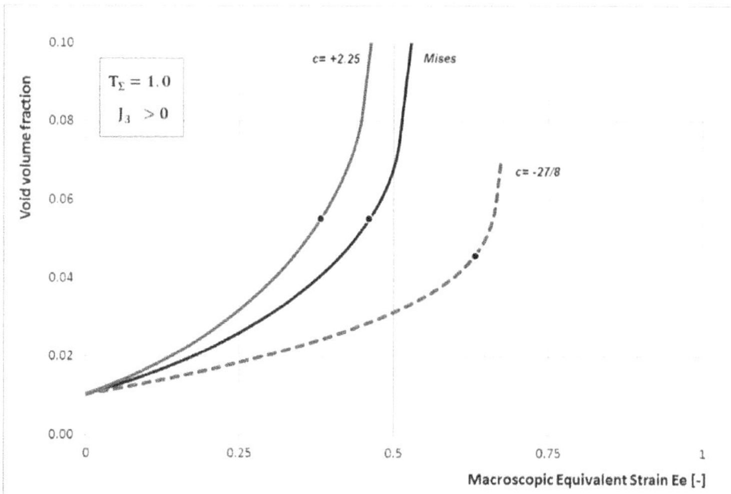

a)

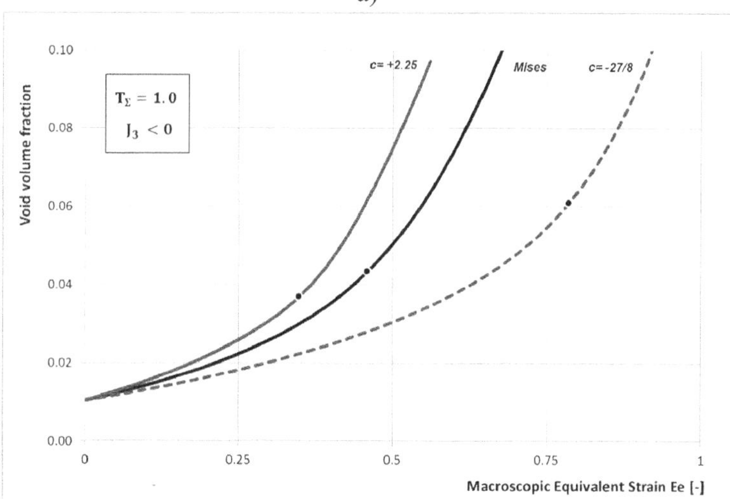

b)

Figure 7.7. *Evolution of the void volume fraction with the macroscopic equivalent strain Ee, obtained by FE cell calculations for an isotropic porous material with a matrix characterized by the Drucker yield criterion corresponding to and c = –27/8, and c = 2.25, respectively in comparison with a Mises material (c = 0, no effect of J_3 in the matrix), obtained for a macroscopic axisymmetric loading corresponding to a fixed triaxiality T_Σ = +1 and (a) J_3 > 0 and (b) J_3 < 0 (black dots denote the onset of void coalescence)*

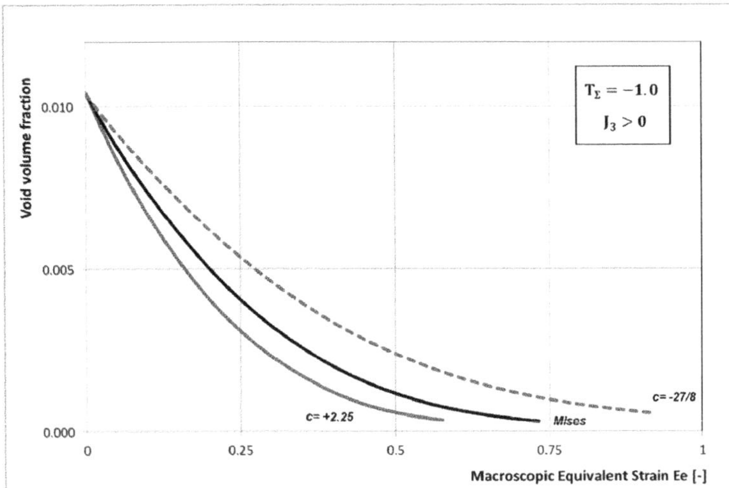

a)

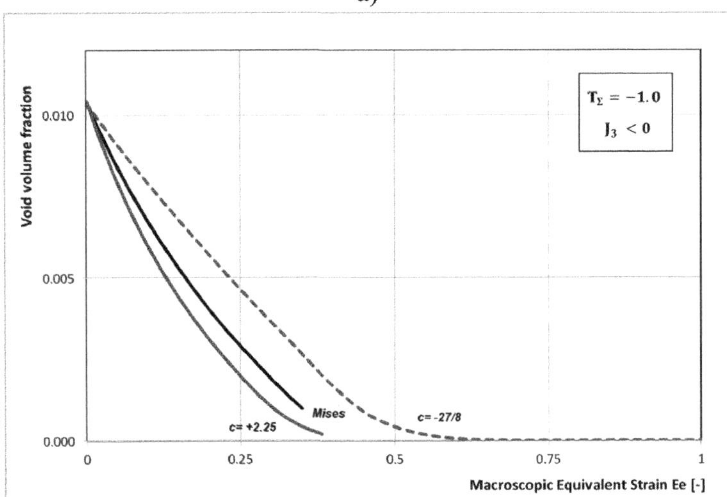

b)

Figure 7.8. *Evolution of the void volume fraction with the macroscopic equivalent strain Ee, obtained by FE cell calculations for an isotropic porous material with a matrix characterized by the Drucker yield criterion corresponding to c = +2.25 and c = –27/8, in comparison with a Mises material (c = 0, no effect of J_3 in the matrix), obtained for a macroscopic axisymmetric loading corresponding to a fixed triaxiality $T_\Sigma = -1$ and (a) $J_3 > 0$ and (b) $J_3 < 0$*

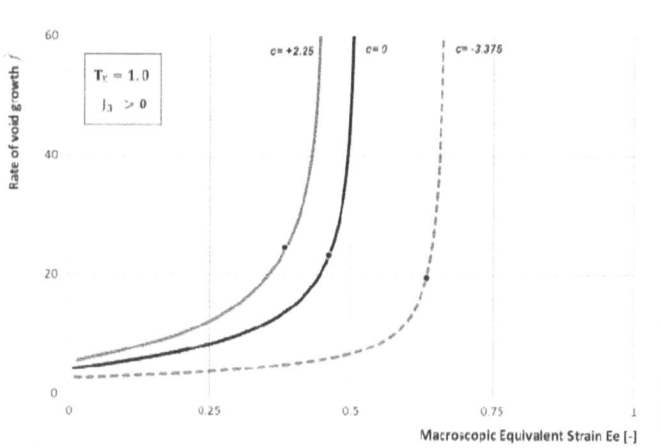

Figure 7.9. *Void growth rate $\dot{f}$ evolution with the macroscopic equivalent strain Ee, obtained by FE cell calculations for porous materials with a matrix characterized by the isotropic Drucker yield criterion c = +2.25 and c = –27/8 (i.e. dependence of J_3 in the matrix) in comparison with a Mises material (c = 0, no dependence of J_3 in the matrix), obtained for a macroscopic axisymmetric loading corresponding to a fixed triaxiality T_Σ = +1 and J_3 > 0 (black dots denote the onset of void coalescence)*

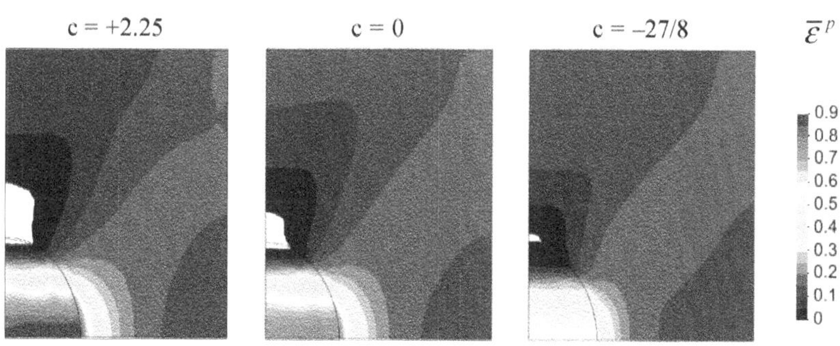

Figure 7.10. *Isocontours of the local effective equivalent plastic strain corresponding to the same value of the macroscopic strain Ee = 0.15 for materials with an isotropic matrix characterized by the Drucker yield criterion with c = +2.25, Mises (c = 0) and c = –27/8, for macroscopic axisymmetric loading corresponding to a fixed triaxiality T_Σ = +1 and J_3 > 0. Note that only in case of material with c = –27/8, almost the entire domain is plastified, which is not the case for the other two materials (elastic domain shown in white). For a color version of the figure, see www.iste.co.uk/brancherie/microstructure.zip*

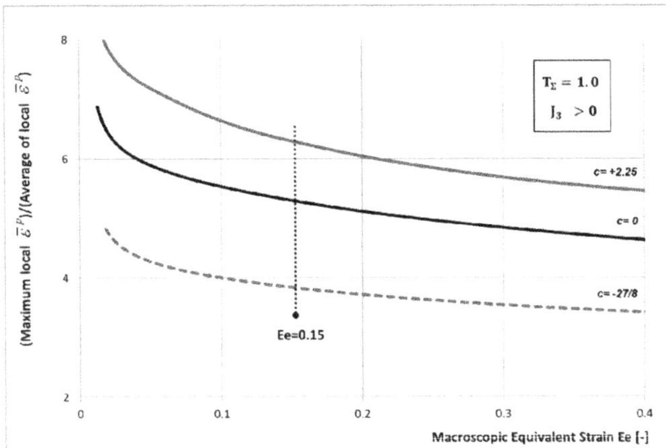

Figure 7.11. *Evolution of the plastic heterogeneity ratio (between the maximum local and average of the local equivalent plastic strain in the matrix) with the macroscopic equivalent strain Ee, obtained by FE cell calculations for materials with an isotropic matrix characterized by the Drucker yield criterion with different contributions of J_3, i.e. c = +2.25 and c = –27/8, in comparison with a Mises material (c = 0), obtained for a macroscopic axisymmetric loading corresponding to a fixed triaxiality T_Σ = +1 and J_3 > 0. Note that, for the same macroscopic deformation, the plastic heterogeneity is markedly different between materials. For a color version of the figure, see www.iste.co.uk/brancherie/microstructure.zip*

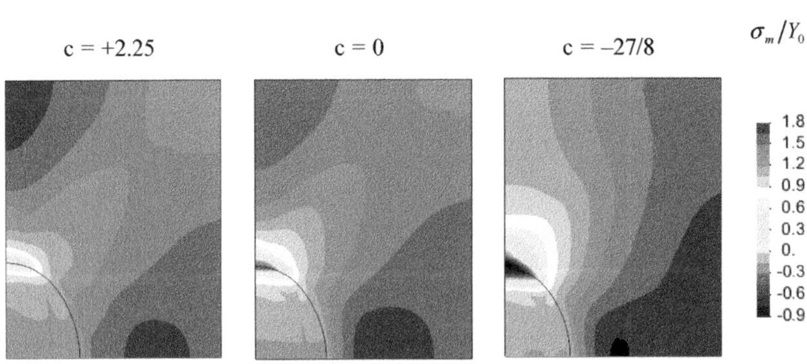

Figure 7.12. *Isocontours of the local normalized mean stress σ_m/Y_0, corresponding to the same value of the macroscopic strain Ee = 0.15 for materials with isotropic matrix characterized by the Drucker yield criterion with c = +2.25, Mises (c = 0) and c = –27/8, for a macroscopic axisymmetric loading corresponding to a fixed triaxiality T_Σ = +1 and J_3 > 0. For a color version of the figure, see www.iste.co.uk/brancherie/microstructure.zip*

The higher the ratio $\bar{\varepsilon}^{\mathrm{p}}_{\max} / \langle \bar{\varepsilon}^{\mathrm{p}} \rangle$, the more heterogeneity there is. For the materials with c = +2.25, c = 0 and c = –27/6, these ratios are 6.3, 5.3 and 3.8, respectively (see Figure 7.11). Moreover, the distribution of the mean stress is markedly different depending on the value of c. Isocontours of the local normalized mean stress σ_m / Y_0 for each material are shown in Figure 7.12. Note that in the porous material with a matrix characterized by c = + 2.25, the local mean stress, σ_m / Y_0, is positive in the entire domain while in the material with c = –27/8, which is almost fully plastified, zones of negative (compressive) mean stress develop near the void. As a consequence, for the latter material, void growth is slowed down when compared with the porous material with a matrix characterized by c = +2.25. Similarly, for loadings at $T_\Sigma = 1$, but at $\mu_\Sigma = -1$ ($J_3^\Sigma < 0$ for the entire loading history), the fact that void growth is fastest in the material with c = 2.25 and slowest in the material with c = –27/8 (Figure 7.7(b)) corroborates with the strong difference in the distribution of the local mean stress shown in Figure 7.13.

In conclusion, although all porous materials were subjected to the same macroscopic tensile axisymmetric loading during the entire deformation process, the specificities of the plastic flow of the matrix, namely how the third invariant affects the plastic deformation of the matrix (which gives the sign of the material parameter c), dramatically affect the local fields. While we have presented in detail the case of tensile loadings at $T_\Sigma = 1$, the same conclusions hold for moderate to high triaxialities. As an example, Figures 7.14 and 7.15 show a comparison between the porosity evolution for the porous materials with a matrix characterized by c = –27/8 and c = 2.25, respectively, for macroscopic loadings corresponding to $T_\Sigma = 0$ and a moderate triaxialiy, $T_\Sigma = 2$.

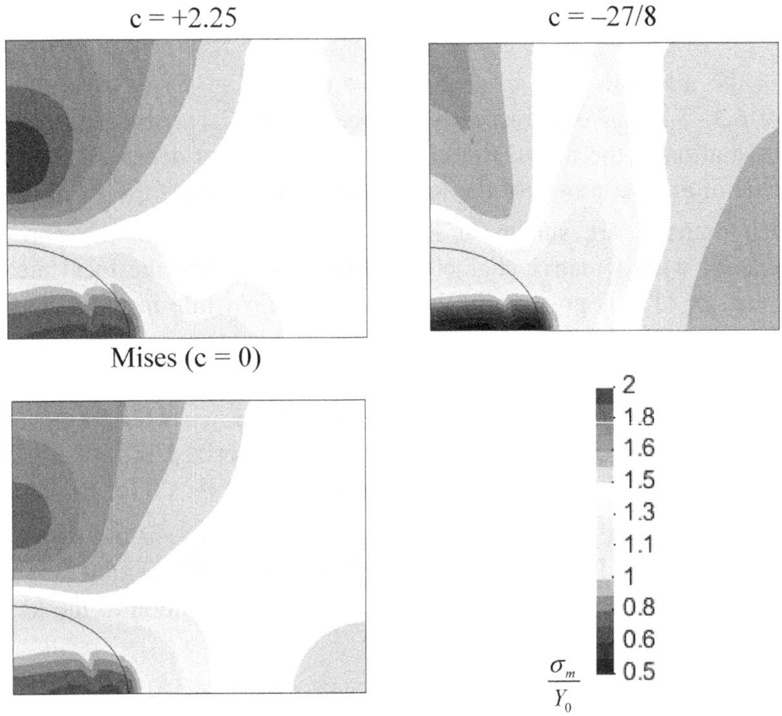

Figure 7.13. *Isocontours of the local normalized mean stress* σ_m/Y_0 *corresponding to the macroscopic strain Ee = 0.20 for materials with isotropic matrix characterized by the Drucker yield criterion with c = +2.25 and c = −27/8, obtained for a macroscopic axisymmetric loading at a fixed triaxiality* T_Σ *= +1 and* $J_3 < 0$*. For a color version of the figure, see www.iste.co.uk/brancherie/microstructure.zip*

7.4. Conclusions

In this work, it was investigated how the deviations in plastic deformation from the von Mises criterion influence the rate at which porosity accumulates in an isotropic porous solid with an incompressible matrix possessing the same response in tension and compression.

For this purpose, FE unit cell calculations were conducted. The plastic deformation of the matrix was considered to be governed by the Drucker yield criterion. This yield criterion is expressed in terms of both invariants of the stress deviator, and involves a parameter c. Depending on the sign of the parameter c, the Drucker yield locus is either interior ($c > 0$) or exterior ($c < 0$) to the von Mises yield locus. If $c = 0$, the Drucker criterion reduces to the von Mises yield criterion. The imposed axisymmetric tensile loadings were such that the principal values of the macroscopic stresses, Σ_1, Σ_2, Σ_3 followed a prescribed proportional loading history corresponding to a specified value of the stress triaxiality. For the same triaxiality, calculations were carried out for loadings at either $J_3^\Sigma > 0$ or $J_3^\Sigma < 0$ (i.e. corresponding to the two possible values of the Lode parameter, $\mu_\Sigma = +1$ or $\mu_\Sigma = -1$, respectively).

It was clearly shown that irrespective of the imposed macroscopic loading, the sensitivity of the plastic deformation to the third invariant of stress, described by the parameter c, has a very strong influence on all aspects of the dilatational response of the porous solids.

Furthermore, a very strong effect of the loading path, in particular the effect of the Lode parameter on void evolution, and ultimately the material's ductility was observed.

While calculations and detailed analysis and comparisons were provided for porous materials with a matrix characterized by $c = -27/8$, $c = 0$ (von Mises) and $c = 2.25$, the same general conclusions can be drawn concerning the differences in dilatational response between materials with a matrix characterized by $c > 0$ and $c < 0$, respectively.

Specifically, irrespective of the loading history (i.e. loadings at $\mu_\Sigma = +1$ or at $\mu_\Sigma = -1$), a porous material with a matrix characterized by $c < 0$ has a slower rate of void evolution than a von Mises material. The latter has a slower rate of void growth than a material with matrix characterized by $c > 0$.

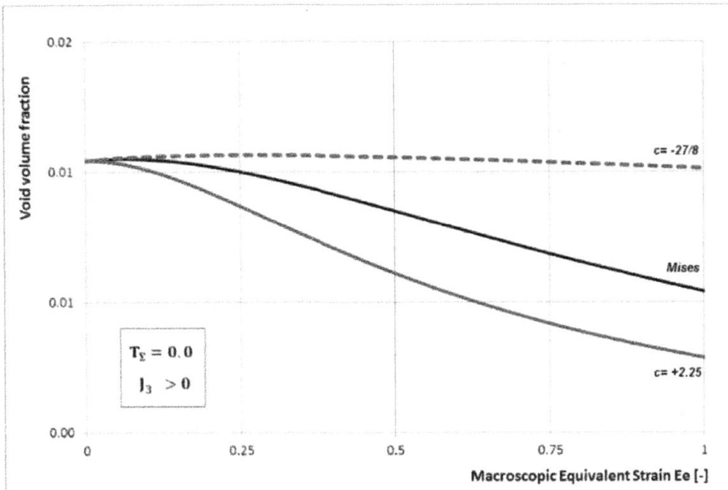

a)

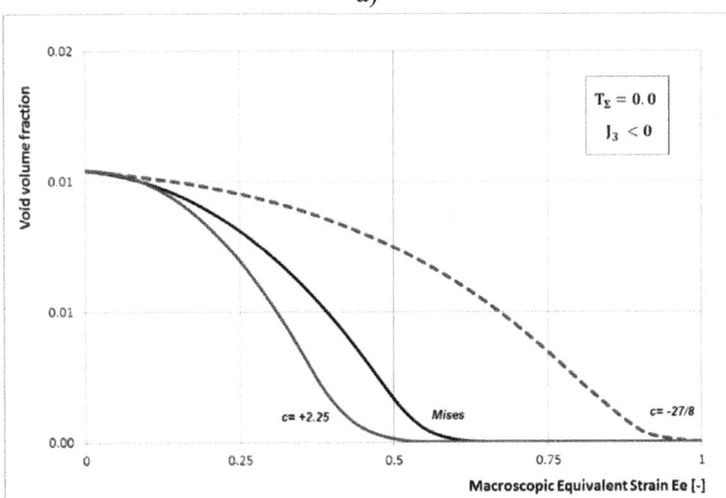

b)

Figure 7.14. *Comparison between the void volume fraction evolution obtained by FE cell calculations for materials with an isotropic matrix characterized by the Drucker yield criterion with c = +2.25, c = 0 (Mises) and c = –27/8, for axisymmetric loadings corresponding to a fixed triaxiality $T_\Sigma = 0$ and (a) $J_3 > 0$ and (b) $J_3 < 0$*

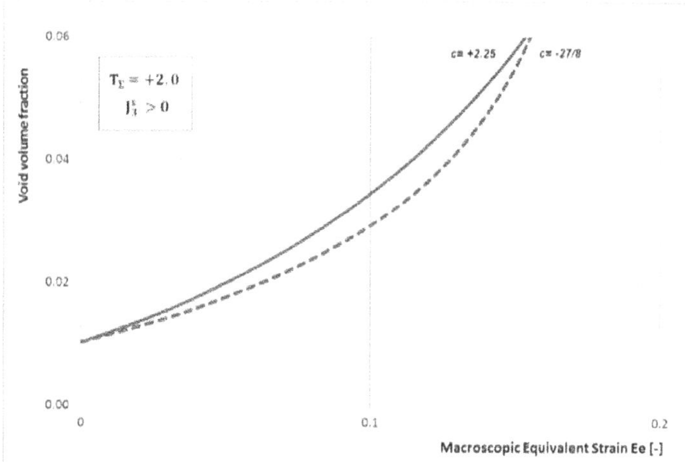

a)

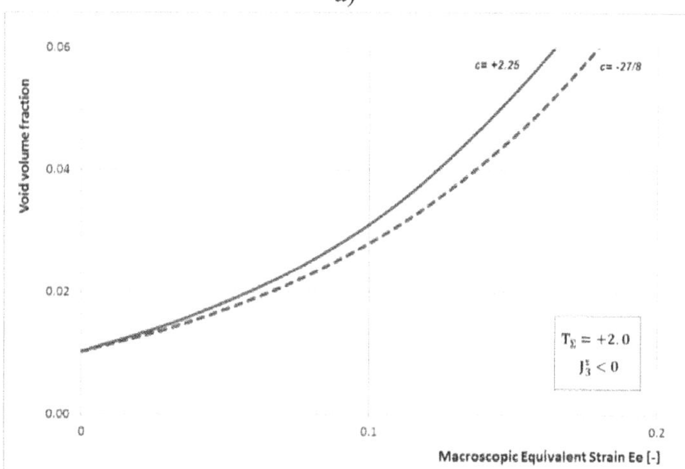

b)

Figure 7.15. *Evolution of the void volume fraction with the macroscopic von Mises equivalent strain Ee, obtained by FE cell calculations for an isotropic material with a matrix characterized by the Drucker yield criterion with c = 2.25 and c = –27/8, for axisymmetric loadings corresponding to a fixed triaxiality of* $T_\Sigma = +2$ *and (a)* $J_3 > 0$ *and (b)* $J_3 < 0$

This is due to the fact that, for $c > 0$, the yield locus of the matrix is interior to that of von Mises, while for $c < 0$, the yield locus of the matrix is exterior to von Mises. Thus, the material characterized by

$c < 0$ dissipates more plastic energy than the von Mises material, and the material characterized by $c > 0$, which in turn results in a slower rate of void growth/collapse. Moreover, on the basis of the results of the unit cell model simulations, it can be concluded that irrespective of the sign of the parameter c, for loadings at positive stress triaxialities, the yield points corresponding to axisymmetric states at $\mu_\Sigma = -1$ are always above those corresponding to at $\mu_\Sigma = -1$ and the rate of void growth is faster for loadings corresponding to $\mu_\Sigma = +1$ (major stress axial) than for loadings for which $\mu_\Sigma = -1$ (major stress is lateral). For loadings at negative stress triaxialities, the reverse holds true. While for a porous von Mises material, the effects of Lode parameter (or J_3^Σ) are due solely to the presence of voids, and are small (see theoretical studies of [CAZ 13, REV 15]), if the plastic flow of the matrix depends on both J_2^Σ and J_3^Σ, the effects of J_3^Σ on the overall response (rate of porosity evolution, ductility) are significant.

7.5. Bibliography

[ALV 14] ALVES J.L., REVIL-BAUDARD B., CAZACU O., "Importance of the coupling between the sign of the mean stress and the third-invariant on the rate of void growth and collapse in porous solids with von Mises matrix", *Modelling and Simulation in Materials Science and Engineering*, vol. 22, no. 2, p. 025005, 2014.

[ALV 15] ALVES J.L., CAZACU O., "Micromechanical study of the dilatational response of porous solids with pressure-insensitive matrix displaying tension-compression asymmetry", *European Journal of Mechanics – A/Solids*, vol. 51, pp. 44–54, 2015.

[CAZ 01] CAZACU O., BARLAT F., "Generalization of Drucker yield criterion to orthotropy", *Mathematics and Mechanics of Solids*, vol. 6, pp. 613–630, 2001.

[CAZ 06] CAZACU O., PLUNKETT B., BARLAT F., "Orthotropic yield criterion for hexagonal closed packed materials", *International Journal of Plasticity*, vol. 22, pp. 1171–1194, 2006.

[CAZ 09] CAZACU O., STEWART J., "Analytic plastic potential for porous aggregates with matrix exhibiting tension-compression asymmetry", *Journal of the Mechanics and Physics of Solids,* vol. 57, pp. 325–341, 2009.

[CAZ 14a] CAZACU O., REVIL-BAUDARD B., CHANDOLA N. *et al.*, "New analytical criterion for porous solids with Tresca matrix under axisymmetric loadings", *International Journal of Solids and Structures*, vol. 51, pp. 861–874, 2014.

[CAZ 14b] CAZACU O., CHANDOLA N., ALVES J.L. *et al.*, "Importance of the consideration of the specificities of local plastic deformation on the response of porous solids with Tresca matrix", *European Journal of Mechanics – A/Solids*, vol. 47, pp. 194–205, 2014.

[DRU 49] DRUCKER D., "Relation of experiments to mathematical theories of plasticity", *Journal of Applied Mechanics*, vol. 16, pp. 349–357, 1949.

[HIL 87] HILL R., "Constitutive dual potentials in classical plasticity", *Journal of the Mechanics and Physics of Solids*, vol. 35, pp. 23–33, 1987.

[KOP 88] KOPLIK J., NEEDLEMAN A., "Void growth and coalescence in porous plastic solids", *International Journal of Solids and Structures*, vol. 24, no. 8, pp. 835–853, 1988.

[LEB 12] LEBENSOHN R.A., CAZACU O., "Effect of single-crystal plastic deformation mechanisms on the dilatational plastic response of porous polycrystals", *International Journal of Solids and Structures*, vol. 49, pp. 3838–3852, 2012.

[LOD 26] LODE W., "Versuche über den Einfluß der mittleren Hauptspannung auf das Fließen der Metalle Eisen, Kupfer und Nickel", *Zeitschrift für Physik*, vol. 36, nos. 11–12, pp. 913–939, 1926.

[LUB 08] LUBLINER J., *Plasticity Theory*, Dover Publications Inc., Mineola, 2008.

[MCC 68] MCCLINTOCK F.A., "A criterion for ductile fracture by the growth of holes", *Journal of Applied Mechanics*, vol. 35, pp. 363–371, 1968.

[MEN 00] MENEZES L.F., TEODOSIU C., "Three-dimensional numerical simulation of the deep-drawing process using solid finite element", *Journal of Materials Processing Technology*, vol. 97, pp. 100–106, 2000.

[MIC 76] MICHNO M.J., FINDLEY W.N., "An historical perspective of yield surface investigations for metals", *International Journal of Non-Linear Mechanics*, vol. 11, pp. 59–82, 1976.

[MIS 13] MISES R.V., "Mechanik der festen Körper im plastisch-deformablen Zustand", *Nachrichten von der Gesellschaft der Wissenschaften zu Göttingen*, vol. Mathematisch-Physikalische Klasse, pp. 582–592, 1913.

[NEE 72] NEEDLEMAN A., "Void growth in an elastic-plastic medium", *Journal of Applied Mechanics*, vol. 39, pp. 964–969, 1972.

[OLI 08] OLIVEIRA M.C., ALVES J.L., MENEZES L.F., "Algorithms and strategies for treatment of large deformation frictional contact in the numerical simulation of deep drawing process", *Archives of Computational Methods in Engineering*, vol. 15, pp. 113–162, 2008.

[REV 13a] REVIL-BAUDARD B., CAZACU O., "On the effect of the matrix tension-compression asymmetry on damage evolution in porous plastic solids", *European Journal of Mechanics – A/Solids*, vol. 37, pp. 35–44, 2013.

[REV 13b] REVIL-BAUDARD B., CAZACU O., THUILLER S. *et al.*, "Effect of stress triaxiality on porosity evolution in notched bars: Quantitative agreement between a recent dilatational model and X-ray tomography data", *Mechanics Research Communications*, vol. 50, pp. 77–82, 2013.

[REV 14] REVIL-BAUDARD B., CAZACU O., "New three-dimensional strain-rate potentials for isotropic porous metals: role of the plastic flow of the matrix", *International Journal of Plasticity*, vol. 60, pp. 101–117, 2014.

[RIC 94] RICHELSEN A.B., TVERGAARD V., "Dilatant plasticity or upper bound estimates for porous ductile solids", *Acta Metallurgica et Materialia*, vol. 42, pp. 2561–2577, 1994.

[SRI 12] SRIVASTAVA A., NEEDLEMAN A., "Porosity evolution in a creeping single crystal", *Modelling and Simulation in Materials Science and Engineering*, vol. 20, pp. 1–23, 2012.

[TAY 31] TAYLOR G.I., QUINNEY H., "The plastic distortion of metals", *Philosophical Transactions of the Royal Society of London, Series A*, vol. 230, pp. 323–362, 1931.

[TRE 64] TRESCA H., "Memoir on the flow of solid bodies under strong pressure", *Comptes-rendus de l'Académie des sciences*, vol. 59, pp. 754–758, 1864.

[TVE 81] TVERGAARD V., "Influence of voids on shear band instabilities under plane strain conditions", *International Journal of Fracture*, vol. 17, pp. 389–407, 1981.

[ZHA 01] ZHANG K.S., BAI J.B., FRANCOIS D., "Numerical analysis of the influence of the Lode parameter on void growth", *International Journal of Solids and Structures*, vol. 38, pp. 5847–5856, 2001.

8

High Performance Data-Driven Multiscale Inverse Constitutive Characterization of Composites

In this chapter, we present recent advances in the multi-scale data-driven inverse constitutive characterization of composites via multi-axial robotic testing, high-performance full-field strain measurement methods and surrogate models. Emphasis is given first on the motivating aspects for multi-scale model inversion for the case of laminated composites, followed by a description of a data-driven inverse constitutive characterization of the bulk lamina of composites via multi-axial loading of laminate-level coupons. A description of a surrogate modeling approach is presented in order to address the computational cost involved in the forward model evaluations associated with the optimization required for material characterization. Finally, we present progress in identifying material properties at the lower scales by utilizing experiments in the laminate-level macro-scale.

Chapter written by John MICHOPOULOS, Athanasios ILIOPOULOS, John HERMANSON, John STEUBEN and Foteini KOMNINELI.

8.1. Introduction

Our main long-term goal is to establish a computational infrastructure that can relate performance and usability specifications of material systems to their optimized manufacturing by design. As described in [MIC 10b], to achieve this goal it is essential to establish the ability to perform inverse analysis on multi-scale cascaded model chains, bridging the macro-scale with the micro- and nano-scales. This is predicated by the fact that behavioral data can be collected more easily in the macro-scale and that manufacturing process parameters (e.g. precursor constituents composition and pressure and temperature of monomer autoclaves) affect the thermo-mechanical properties of these material systems at the micro length scales, as shown elsewhere [FU 16]. That is, in order to achieve desired functional outcomes at the macro-scale by controlling the process parameters in the lower length scales, it is necessary to connect the various models that can be built and linked across the relevant scales. This process is depicted in Figure 8.1. Specifically, instead of determining the proper process parameters from the functional specification of the material system to be developed through direct inversion of a single model (as shown in the top path of Figure 8.1), we envision the inversion of a cascade of models distributed along various scales through the lower path of the figure.

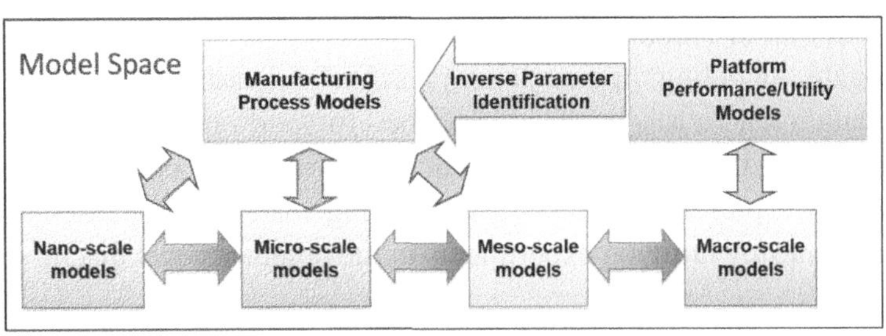

Figure 8.1. *Computational framework depicting equivalency of inverse problems connecting macro-scale functional specification with process parameters with inverse problem of cascaded chain of models along multiple length scales*

Our main short-term goal is to demonstrate that such a framework can be useful for designing a material system relevant to a wide variety of applications. A material system that is suitable to demonstrate at least a three-level length-scale cascade is that of the polymer matrix fiber composites (PMFCs). At the micro-scale, the constituents of the composite, namely the fibers and the matrix, appear as individual deformable continua with distinctly different properties. At the meso-scale, the thin individual lamina (or ply) can be abstracted as a homogenized transversely isotropic material sub-system with a major axis of orthotropy along the direction of the fibers. At the macro-scale, stacks of laminae form the composite material laminates that in general appear as anisotropic materials, but in some cases can appear as orthotropic or even transversely isotropic materials depending on the stacking sequence.

In order to achieve these goals, and because mechanical testing can occur easily at the macro-scale, we have decided to examine the potential of solving the composition of the structural equilibrium equations at each scale in such a manner that instead of determining the linear anisotropic elastic parameters of the laminate at the macro-scale level as an effective material, we determine the properties of the individual constituents at the micro-scale. This has a particularly important utility especially if we consider that some of the anisotropic moduli of the fibers cannot always be determined experimentally.

In addition, we are motivated by a specific variation of our long-term goal to be able to assess performance specification requirements that sometimes involve quantification of material aging in terms of damage mechanisms like diffused resin micro-cracking, fiber–matrix interface degradation or fiber breaking. In this effort, it is important to be able to identify quantified forms of these damage mechanisms and associated morphologies that also manifest in lower scales, by performing experiments in the macro-scale. Therefore, it is of essence to demonstrate the ability of a methodology to identify lower-scale entities such as the potentially degraded elastic properties of the interface between fibers and matrix.

The progress in the areas of computers and information in engineering and in mechatronic automation has enabled us to pursue the research areas of multi-axial robotic testing and design optimization for the purpose of data-driven constitutive characterization of anisotropic material systems. Although our efforts began three decades ago, here we present our progress, as it relates to the multi-scale inversion described above. This methodology is motivated by the data-driven requirements of employing design optimization principles for determining the constitutive behavior of composite materials as described in our recent work [MIC 11a, MIC 11c]

Traditionally, the constitutive characterization of composite materials has been determined through conventional uniaxial tests, mainly aiming at the estimation of the elastic properties for simplified stress–strain states. Typically, extraction of these properties involves several uniaxial tests conducted with specimens mounted on uniaxial testing machines, where the major orthotropic axis of any given specimen is angled relative to the loading direction. In addition, specimens are designed such that a homogeneous state of strain is developed over a well-defined area, which is required for the purpose of measuring stresses and strains through the measurement of the respective reaction forces and displacements [DAN 05, CAR 02]. Consequently, the use of uniaxial testing machines imposes requirements of using multiple specimens, griping fixtures and multiple experiments without the option of studying multi-axial effects in their full range. The requirement of a homogeneous state of strain frequently imposes restrictions on the sizes and shapes of specimens to be tested. These requirements result in increased cost and time and inefficient characterization processes.

To address these issues and to extend the characterization to a multi-axial state of strain in both the linear and nonlinear regimes, multi-degree of freedom-automated mechatronic testing machines were introduced at the Naval Research Laboratory (NRL), in conjunction with energy-based inverse characterization methodologies, [MAS 92, MAS 95]. This development was the first of its kind and has continued ever since [MIC 03, MIC 04, MIC 08]. Thus, our approach employs mechatronic testing, and although it enables multi-axial loading that induces inhomogeneous states of strain, it still requires multiple specimens. However, these specimens are tested in an automated manner with high testing throughput that has reached a peak rate of 28 specimens per hour.

In addition, the development of flexible full-field displacement and strain measurement methods during mechanical tests has afforded the opportunity of alternative characterization methodologies [PIE 07, COO 07, MIC 09b, ILI 13, ILI 14]. Methods such as Moire and speckle interferometry, digital image correlation (DIC) and some high-performance methods, such as the meshless random grid (MRG) method and the direct strain imaging (DSI) method, have been used mostly for elastic characterization of various materials [MIC 09b, BRU 02, SCH 03, KAJ 04, ILI 16]. The resulting measurements are used for the identification of constitutive model constants, via the solution of an appropriately formed inverse problem, using various computational techniques. In particular, not only the authors [MAS 92, MAS 95, MIC 03, MIC 04, MIC 08, MIC 09b] but also other groups [KAJ 04, MEU 98, MOL 05] have focused on developing a mixed numerical/experimental method that identifies the material's elastic and inelastic constants by minimizing an objective function formed by the difference between full-field experimental measurements and the corresponding analytical model predictions via an optimization method.

Our approaches are based mostly on energy conservation and can thus be classified according to computational cost in relation to the repetitive use of finite-element analysis (FEA) or not. It is important to clarify that digitally acquired images are processed by software [MIC 09a] that implements the meshless random grid (MRG) method [MIC 09b, AND 06, ILI 12, ILI 09a] to extract the displacement and strain field measurements as well as the boundary displacements required for material characterization. The development and use of the MRG method was motivated by the requirements of resolution, accuracy and computational efficiency that were higher than those available from other full-field methods. Reaction forces and redundant boundary displacement data are acquired from displacement and force sensors integrated with NRL's 6-degrees of freedom (6-DoF) multi-axial loader called NRL66.3 [MIC 10a]. In an effort to address the computational cost of the FEA-in-the-loop approaches, the authors have initiated a dissipated and total strain energy density determination approach that has recently been extended to a framework that is derived from the total potential energy and the energy conservation, which can be applied directly with full-field strain measurement for characterization [MAN 08, MIC 09c].

In the next section, we present a description of the 6-DoF loader, the associated loading space characteristics and the experimental campaign. The following section introduces an overview of two energy-based constitutive formalisms necessary for the characterization. The implementation details of the design optimization methodology follow along with characterization and validation results. The subsequent section presents the utilization of surrogate models for improving the computational efficiency of the optimization loop required for the constitutive characterization. The following section presents an application of the multi-scale cascaded inversion framework that enables utilization of experimental data at the macro-scale for determining properties at the micro-scale of the constituents. Finally, the chapter ends with a description of the conclusions drawn.

8.2. Automated multi-axial testing

NRL66.3 is a custom-designed and prototyped system that represents the third generation of 6-DoF loaders built by our group for the purpose of automating the process of multi-axial testing required for the data-driven constitutive characterization of composites. It was conceived to ensure that the material to be characterized will be exposed to a very comprehensive set of deformation states such that the richest set of strain states will develop inside this material. The system has a recursive character because its characteristic hexapod configuration is repeated six times and is described in [MIC 10a] and shown in Figure 8.2.

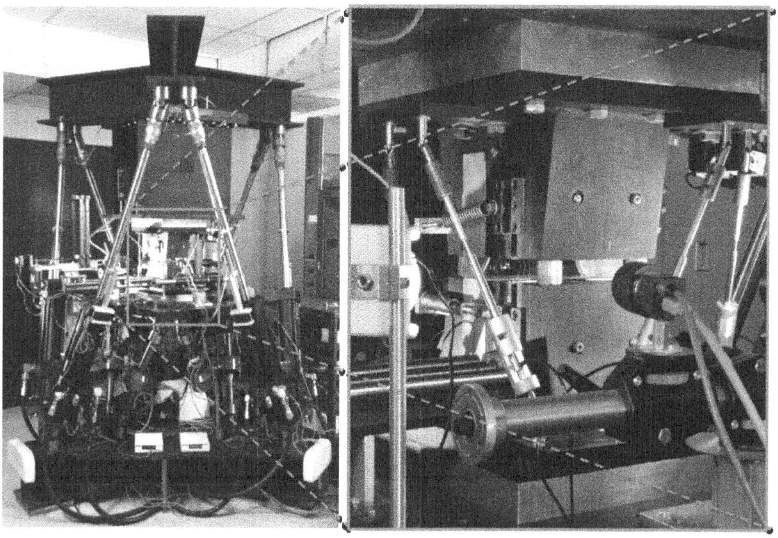

Figure 8.2. *NRL66.3: 6-DoF mechatronically automated system for the multi-axial testing of composite materials: long (left) and near (right) views of the system*

The typical flat specimen geometry is shown in Figure 8.3, where the dark gray areas indicate the part of the specimen that is placed into the grips and the light gray area represents the part that deforms under the influence of the applied load. The double notches on the specimen are introduced to disturb the strain field and to ensure that some areas of the specimen (not necessarily near the notch roots) will experience nonlinear constitutive response due to the corresponding strain fields. The dimensions are in "millimeters (inches)". The white dots on the deformation domain as shown in Figure 8.3 (right side) are markers required for measuring the displacement and strain field components via the MRG method [ILI 09a, ILI 09b, MIC 09a].

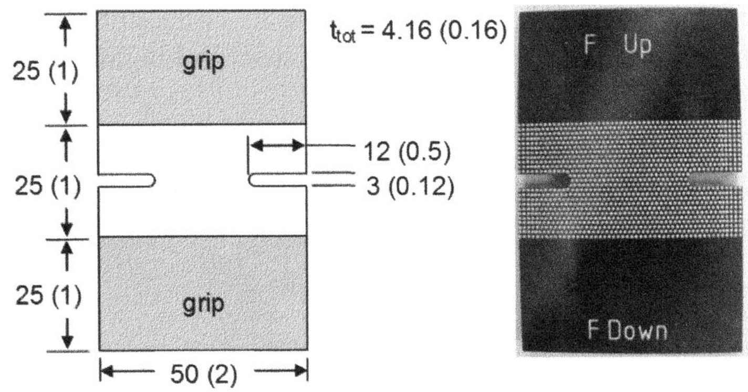

Figure 8.3. *Typical specimen schematic (left) and photographic view (right) of prepared specimen*

8.2.1. *Loading space*

Because the main mission of the system is to induce a set of strain states that is as complete as possible, it is designed such that it can be programmed to follow displacement- and rotation-controlled loading paths while it can measure boundary displacements and reaction forces and moments resulting from the mechanical response of the specimens under testing.

The hexapod configuration effectively converts the linear extension motions of the six actuators to the three translations and the three rotations imposed by the upper grip on the specimen. These three translations and three rotations effectively define a six-dimensional space that can be sampled by commanding NRL66.3 to follow radially proportional paths and acquire data along these paths. The term proportional path denotes that every path i, when represented as a vector starting in the origin and ending at point j, can be resolved to its vector space basis components in the 6-D frame $(\mathbf{u}_x, \mathbf{u}_y, \mathbf{u}_z, \mathbf{r}_x, \mathbf{r}_y, \mathbf{r}_z)$ according to:

$$\mathbf{p}_i^j = p_{i1}^j \mathbf{u}_x + p_{i2}^j \mathbf{u}_y + p_{i3}^j \mathbf{u}_z + p_{i4}^j \mathbf{r}_x + p_{i5}^j \mathbf{r}_y + p_{i6}^j \mathbf{r}_z, \qquad [8.1]$$

where $p_{i1}^j, p_{i2}^j, p_{i3}^j, p_{i4}^j, p_{i5}^j, p_{i6}^j$ are the magnitudes of each component, and furthermore:

$$p_{ik}^j = r_i^j a_{ik}^j, k \in \{1,2,3,4,5,6\}, \text{ with } a_{ik}^j = a_{ik}^{j+1}. \qquad [8.2]$$

A convenient parameterization of the loading paths can be obtained by invoking a transformation from the hyperspherical to its Cartesian frame of reference as follows:

$$\left.\begin{array}{l} p_{i1}^j = r_i^j \cos(\phi_1^i) \\ p_{ik}^j = r_i^j \cos(\phi_{k-1}^i) \prod_{m=1}^{k-1} \sin(\phi_m^i) \text{ for } k = 2, ..., 6 \end{array}\right\}, \qquad [8.3]$$

where r_i^j and ϕ_k^j are the magnitude of the path and the angles between it and the planes defined by the basis vectors of the 6-D kinematic space.

At least as many specimens as the number of paths were used to sample the material system (material coupon) excitation space (i.e. the loading space).

The number of paths as a function of the dimension n of the space and the number of equatorial points n_{ep} is given by [MIC 08]:

$$n_{paths} = (4 + 4n_{ep})(1 + 2n_{ep})^{n-2}. \qquad [8.4]$$

An intermediate scenario (in terms of required number of specimens) could be based on the fact that the rotation components generate motion on the plane they are normal to. Thus, only four out of the six components can be linearly independent. In this case, $n = 4$ and for $n_{ep} = 1$, equation [8.4] generates 72 paths. This promotes the solution of utilizing one of the fifteen 4-D sub-spaces that can be used for defining the loading paths.

A computationally intensive methodology for determining the best sub-space is presented in detail elsewhere [ILI 10]. An appropriate metric representation with respect to strain state volumetric histograms was developed, and the notion of strain-state clouds (SSCs) [ILI 10] was introduced as a means of visualizing these histograms. Extensive numerical experiments were conducted on the basis of FEA computations of the elastic response of a composite specimen with edge notches. They were used as the means for generating all possible strain states generated by loading paths in the corresponding sub-spaces. A typical example of the relevant SSCs is depicted in Figs.8.4a, 8.4b for loading case 02 (u_y, r_x, r_y, r_z). The color scale from blue to red indicates the distance from $\{0, 0, 0\}$ and is only used as a visual aid assisting qualitative comparison.

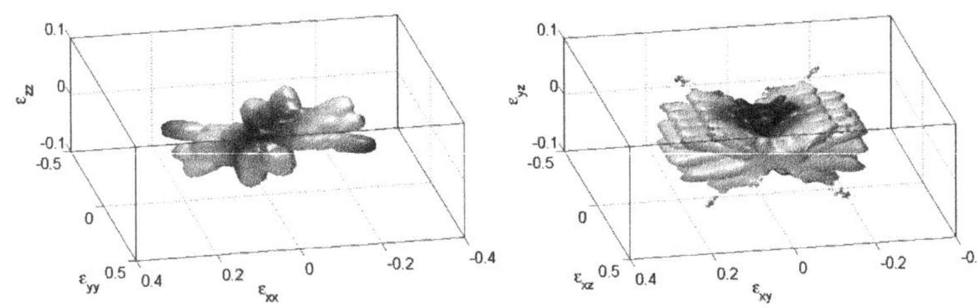

(a) Normal for L.C. 2 (u_y, r_x, r_y, r_z) (b) Shear for L.C. 2 (u_y, r_x, r_y, r_z)

Figure 8.4. *Normal and Shear Strain State Clouds for a specimen with stacking sequence* $[-30°, 30°]_{16}$. *For a color version of the figure, see www.iste.co.uk/brancherie/microstructure.zip*

The ratio of the volume of each SSC to the sum of the volumes of all SSCs created by all possible loading paths within all sub-spaces was used as the load sub-space ranking metric. It was established that the sub-spaces involving all three rotations and one of the two in-plane translations generate the richest strain state sets. Here, the term "richest" signifies that the 4-D sub-spaces contain the largest sets of strain states that also appear in the full 6-D dimensional space. On the basis of an energy loss criterion, the final sub-space selected was that involving the three rotations and the tension–compression loading case.

8.2.2. *Experimental campaign*

The lowest number of specimens used on the basis of economical considerations was determined using a single specimen per loading path. For reasons of simple redundancy, we decided to repeat each path, and therefore each load path was followed twice by testing two specimens of identical material systems. Thus, 144 specimens per material system are required. Each material system was defined to be a balanced laminate with an alternating angle of fiber inclination per ply relative to the vertical axis of the specimen that is perpendicular to the axis defined by the two notches of the specimen. Four angles were used, $+/- 15, +/- 30, +/- 60, +/- 75$ degrees, and therefore $4 \times 144 = 576$ specimens were manufactured. The actual material used to make the specimens was an AS4/3501-6 carbon fiber/epoxy resin system.

However, because we needed to study ply thickness length scale effects, we manufactured an equal amount of specimens, where the cross-plied unidirectional laminas were made with four plies each, thus leading to thicker uniaxial laminas. This doubled the number of total specimens to 1,152, and this was the final number of required tests. All specimens were tested using NRL66.3 in May 2011. A very small sample of the testing process is shown in a video which can be found at URL http://www.youtube.com/watch?v=Cp18y3HAqsM. On May 13, 2011, the system achieved a throughput of 20 tests/h for a total of 132 tests. On May 27, the system achieved a throughput of 25 tests/h for a total of 216 tests. On June 15, 2011, the system reached its peak throughput of 28 tests/h. All 1,152 tests were completed in 12 working days. The test yielded 13 TB of data from the sensors and the cameras of the system.

8.3. Constitutive formalisms

To derive the mathematical constitutive form of an anisotropic material, we adopted a modified anisotropic strain energy density function as an additive decomposition of both an elastic and an inelastic (which accounts for damage) part. However, certain classes of composite materials reach failure after small strains and some under large strains. For this reason, we adopted two cases, one involving a small (infinitesimal) strain formulation (SSF) and another involving a finite (large) strain formulation (FSF).

8.3.1. *Small strain formulation*

For the SSF, we introduce a strain energy density (SED) function that, in its most general form, can be represented as a scaled Taylor expansion of the Helmholtz free energy of a deformable body, which is in terms of small strain invariants of the form [MIC 14]:

$$U_{SSF}(I_n) = s_0 + s_{11}I_1 + \frac{1}{2}s_{12}I_1^2 + \frac{1}{3!}s_{13}I_1^3 + \frac{1}{4!}s_{14}I_1^4 + \ldots, \quad [8.5]$$

where $I_n; n = 1, 2, 3$ are the strain tensor invariants.

After an additive decomposition of the SED in terms of recoverable and irrecoverable parts, its differentiation with respect to strain components to derive the constitutive law and subsequent algebraic manipulations [MIC 14], we can derive the expression:

$$\begin{aligned} U_{SSF} &= U^R_{SSF}(S; \varepsilon_{ij}) + U^I_{SSF}(D; \varepsilon_{ij}) = \\ &= \frac{1}{2}s_{ijkl}\varepsilon_{ij}\varepsilon_{kl} - s_{ijkl}\frac{1}{e(2+p_{ij})p_{ij}q_{ij}^{p_{ij}}}\varepsilon_{ij}^{1+p_{ij}}\varepsilon_{kl}. \end{aligned} \quad [8.6]$$

Employing the Voigt [DAN 05] notation for the case of a general orthotropic material yields the constitutive relation [MIC 14]:

$$\begin{bmatrix} \sigma_{xx} \\ \sigma_{yy} \\ \sigma_{zz} \\ \sigma_{xz} \\ \sigma_{yz} \\ \sigma_{xy} \end{bmatrix} = \begin{bmatrix} \breve{s}_{xx} & s_{xy} & s_{xz} & 0 & 0 & 0 \\ s_{xy} & \breve{s}_{yy} & s_{yz} & 0 & 0 & 0 \\ s_{xz} & s_{yz} & \breve{s}_{zz} & 0 & 0 & 0 \\ 0 & 0 & 0 & \breve{s}_{xz} & 0 & 0 \\ 0 & 0 & 0 & 0 & \breve{s}_{yz} & 0 \\ 0 & 0 & 0 & 0 & 0 & \breve{s}_{xy} \end{bmatrix} \begin{bmatrix} \epsilon_{xx} \\ \epsilon_{yy} \\ \epsilon_{zz} \\ \epsilon_{xz} \\ \epsilon_{yz} \\ \epsilon_{xy} \end{bmatrix}, \quad [8.7]$$

where:

$$\breve{s_{ij}} = s_{ij}\left(1 - \bar{d_{ij}}\right) \quad [8.8]$$

and:

$$\bar{d_{ij}} = \frac{1}{ep_{ij}q_{ij}^{p_{ij}}}\varepsilon_{ij}^{1+p_{ij}}. \quad [8.9]$$

For a transversely isotropic material, the number of material parameters is $5 + 10 = 15$ for a 3-D state of strain and $4 + 8 = 12$ for a plane stress state [JU 98, MIC 14].

8.3.2. *Finite strain formulation*

The FSF can be written in a form expressing the volumetric versus distortional decomposition similar to the expression given by [KAL 00]:

$$U_{FSF} = (1 - d_v)W_v(J) + (1 - d_d)W_d(\bar{C}, A \otimes A, B \otimes B), \qquad [8.10]$$

with the damage parameters $d_k \in [0, 1], k \in [v, d]$ defined as:

$$d_k = d_{ka}^{\infty}\left[1 - e^{\left(-\frac{a_k(t)}{\eta_{ka}}\right)}\right], \qquad [8.11]$$

where $a_k(t) = \max\limits_{s \in [0,t]} W_k^o(s)$ is the maximum-energy component reached so far and $d_{ka}^{\infty}, \eta_{ka}$ are two pairs of parameters controlling the energy dissipation characteristics of the two components of SED. In this formulation, $J = \det \underline{F}$ is the determinant of the deformation gradient, $\bar{C} = \underline{F}^T\underline{F}$ is the right Cauchy–Green (Green deformation) tensor, A, B are constitutive material directions in the undeformed configuration and $A \otimes A, B \otimes B$ are microstructure structural tensors expressing fiber directions.

Under the FSF, the material characterization problem involves determining the 36 coefficients (at most) of all monomials in addition to the compressibility constant d and the four parameters used in equation [8.11]. It follows that there can potentially be a total of 41 material constants [MIC 14].

8.4. Meshless random grid method for experimental evaluation of strain fields

As will be clear in the following section, it is essential that the strain tensor component fields are evaluated experimentally in order to use the relevant data for the design optimization methodology described in the next section.

After assessing the flexibility, simplicity, sensitivity, accuracy and computational efficiency requirements needed for this effort, it was concluded that existing methods did not satisfy these requirements to our satisfaction. For this reason, we developed the MRG method. Detailed descriptions of the method can be found in [MIC 11b, ILI 14]. Here, we present a short overview for the sake of completeness and we urge the reader to seek more details elsewhere [ILI 14].

The MRG method is based on the meshless representation of the displacement field considering the values of the displacement on the nodes, while the strain tensor is identified by the differentiation of those displacements.

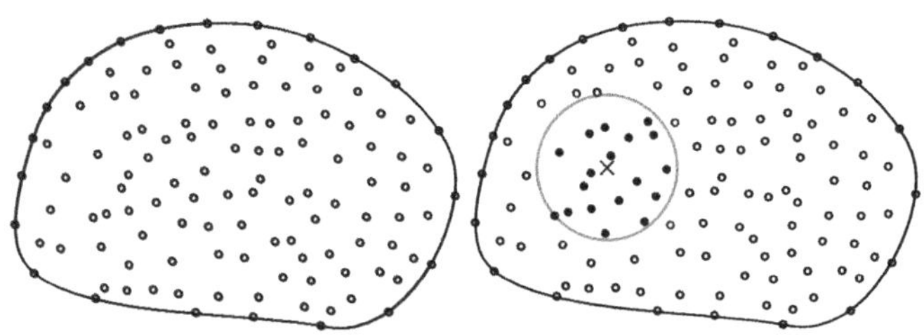

(a) Nodes in a domain

(b) A Circular domain of support for a point of interest

Figure 8.5. *Meshless representation*

In order to accomplish full-field measurement of strain fields, the MRG method requires a procedure involving the following four steps [ILI 07, ILI 14]:

1) A specimen is prepared by marking it with a random distribution of black dots at the area of interest.

2) Images of the non-deformed and deformed specimens are acquired, and a simple labeling algorithm is used to identify the centroids (centers of the marked dots) on each image.

3) A point-matching algorithm identifies the correspondence between centroids of the dots between the two images and calculates the respective displacement.

4) The centroid displacement values previously obtained are used to calculate the full-field values of displacement and strain through meshless functions. In the last step, there are various choices from typical meshless functions [LIU 09], including the moving least-square (MLS) method for shape function construction and point interpolation (PI) methods. In the present work, the shape function is constructed through the MLS method.

The MLS approximates the displacement field using an expression of the form:

$$u^h(\mathbf{x}) = \mathbf{\Phi}(\mathbf{x})\,\mathbf{U}_S, \quad [8.12]$$

where $\mathbf{U}_S$ is a vector of displacement components of the nodes in the vicinity of the point represented by the vector $\mathbf{x}$, while:

$$\mathbf{\Phi}(\mathbf{x}) = [\phi_1(\mathbf{x}), \phi_2(\mathbf{x}), ..., \phi_n(\mathbf{x})], \quad [8.13]$$

are shape functions calculated from expressions of the form,

$$\phi_I(\mathbf{x}) = \sum_j^m p_j(\mathbf{x}) \left[\mathbf{A}^{-1}(\mathbf{x})\,\mathbf{B}(\mathbf{x})\right]_{jI} = \mathbf{p}^T\mathbf{A}^{-1}\mathbf{B}_I, \quad [8.14]$$

where p_j represent the components of the polynomial basis $\mathbf{p}^T$ and $\mathbf{A}, \mathbf{B}_I$ are matrix compositions of the polynomial basis and appropriate weight functions. The displacement field expressed by equation [8.12] is subsequently differentiated to provide the surface strain components, fields. More details can be found in [ILI 14].

8.5. Inverse determination of HDM via design optimization

In order to determine the material parameters of the constitutive model that represents the behavior of the bulk lamina, the inverse problem at hand was solved through a design optimization approach described in more detail elsewhere [MIC 14]. The optimization process outline is shown in Figure 8.6.

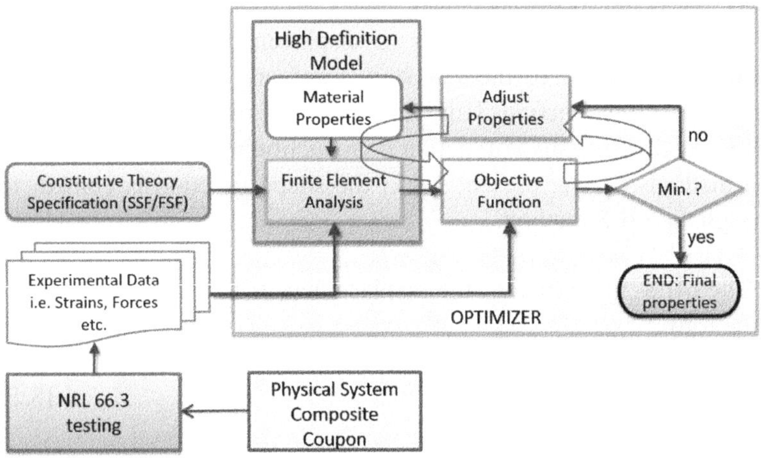

Figure 8.6. *Design optimization data flow for inverse characterization via utilization of HDM*

This particular scheme exercises the optimizer for the case of the high-definition model (HDM) that represents the full FEA model for the test coupon geometry. Details of this model are presented in [MIC 11c, MIC 11a] due to space limitations.

Two objective functions were constructed, and both utilized the fact that through the REMDIS-3D software [MIC 11b] developed by our group, we can obtain full-field measurements of the displacement and strain fields over any deformable body as an extension of the REMDIS-2D software that is based on the MRG method [MIC 09b, AND 06, ILI 12, ILI 09a, ILI 14, ILI 14]. Thus, our experimental measurements for the formation of the objective functions were chosen to be the strains at the nodal points of the FEM discretization. The first objective function chosen was based entirely on strains and is given by:

$$J^{\varepsilon} = \sum_{k=1}^{N} \left(\sum_{i=1}^{2} \sum_{j=i}^{2} \left(\left[\varepsilon_{ij}^{\exp} \right]_k - \left[\varepsilon_{ij}^{fem} \right]_k \right) \right)^2, \qquad [8.15]$$

the second objective function is given in terms of surface strain energy density according to:

$$J^{U} \approx \oint_{\partial\Omega} \left(U^{\exp} - U^{fem} \right)^2 dS. \qquad [8.16]$$

The quantities $U^{\exp}$ and U^{fem} are the surface strain energy densities formulated using the experimental strains and the FEM-produced strains, respectively. In the last two equations, the quantities ε_{ij}^{fem} and U^{fem} require an initial guess of the material properties as the coefficients of the respective monomials in the strain energy density function representation. Clearly, as the optimization progresses, the values of these parameters are modified until they converge to their final values.

For the case of the FSF, a two-stage optimization was performed. In the first stage, we determined the values of the coefficients of the strain invariant monomials such that the FSF matches the SSF by constructing and minimizing an objective function of the form:

$$J^{URF}(d, a_1, b_1, c_1, d_1) \approx \oint_{\partial\Omega} \left(U_{FSF}^{R} - U_{SSF}^{R} \right)^2 dS \qquad [8.17]$$

This was done in order to establish the proper parameters of the FSF model that matches the SSF model.

In the second stage, the material parameters encoding the damage behavior of the FSF were determined through the minimization of the objective function:

$$J^{UF}(d, a_1, b_1, c_1, d_1, d_a^{\infty}, \eta_a) \approx \oint_{\partial\Omega} \left(U_{FSF}^{1} - U_{FSF}^{2} \right)^2 dS \qquad [8.18]$$

where $U^1_{FSF}(\bar{d}, \bar{a}_1, \bar{b}_1, \bar{c}_1, \bar{d}_1, \bar{d}^\infty_a, \bar{\eta}_a)$ is the SED of the FSF for the parameters $\bar{d}, \bar{a}_1, \bar{b}_1, \bar{c}_1, \bar{d}_1$ computed from the previous step of matching the FSF model with the SSF model. The unknown constitutive model was chosen to be represented by another FSF SED, namely $U^2_{FSF}(d, a_1, b_1, c_1, d_1, d^\infty_a, \eta_a)$.

8.5.1. *Numerical results of design optimization*

For the purpose of demonstrating numerically the effectiveness of the aforementioned design optimization concepts, the material selected for generating the necessary simulated experimental data is a typical laminate constructed from a carbon fiber/epoxy resin lamina system of type AS4/3501-6. As can be evidenced by the reported values of the elastic constants for this material system by several sources [DAN 05, RYA 90, WEG 00, VAI 97] and as summarized in [MIC 14], the variability of the reported values extends from 11.1% to 78.2%. It is therefore important to identify the set of elastic material properties before and after a batch of new materials is manufactured or before a material system is used for design, material qualification or material certification.

To demonstrate the applicability of the proposed approach of solving the inverse problem for the SSF with the proposed design optimization method, we present here an example of using real data from a multi-axially loaded specimen from a test conducted by utilizing NRL66.3. The model characteristics of the specimen used are presented in Figure 8.7, in terms of both the discretization model and potential boundary conditions, and a detail in the area of the left notch shows a stacking of $[+60, -60]16$ with each lamina made out of AS4/3501-6.

By applying the outlined optimization approach for the case of the SSF, we identified the elastic constants as shown in Table 8.1, in comparison to those of [DAN 05].

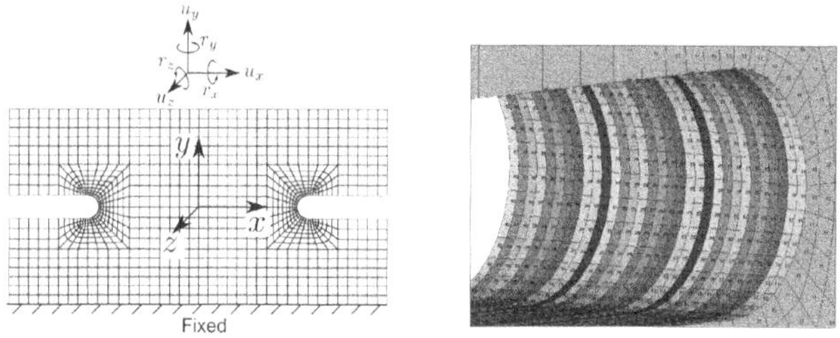

(a) Discretization and boundary conditions of typical specimen used for material characterization

(b) Detail of FEA model near one of the notches

Figure 8.7. *Finite-element model*

Ref	E_{11} [GPa]	E_{22} [GPa]	ν_{12}	ν_{23}	G_{12}[GPa]
Daniels [DAN 05]	147.0	10.3	0.27	0.28	7.0
Present	125.0	10.8	0.27	0.32	7.96

Table 8.1. *Engineering properties of AS4/3501-6 laminae*

By performing FEA for the cases that correspond to the specific loading path corresponding to an experiment, we can now compare the predicted distribution of any component of the strain or stress tensor. In order to demonstrate how well the FSF formulation can capture the behavior of the characterization coupons used to obtain the data used in the characterization process, we present a typical example in terms of the distributions of ε_{yy} as measured by the MRG method (left column) and as predicted by the FSF theory (right column), for both the front (top row) and the back (bottom row) in Figure 8.8.

More examples depicting comparisons between full-field measurements and model-predicted results for additional loading paths can be found in [MIC 14].

To determine the applicability of the constitutive characterization beyond the coupon geometry and length scale, additional validation tests have been performed on structures of shapes and layups different to those of the characterization coupons, to perform validation comparisons. A discussion of this activity is briefly presented in a more extended form in [MIC 12a].

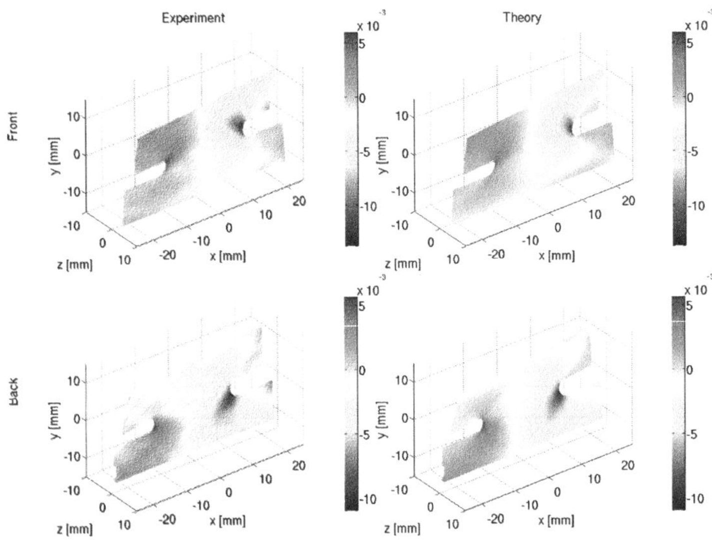

Figure 8.8. *Comparison of (ε_{yy}) field between measured (left) and the identified model using the FSF for the case of in-plane rotation and torsion of a $+/-60$ degree laminate. For a color version of the figure, see www.iste.co.uk/brancherie/microstructure.zip*

8.6. Surrogate models for characterization

While the previously described methodology works well from the perspective of determining the unknown material parameters, the FEA model is evaluated at every iteration of the optimization, indicating that the computational cost may be impracticable [MIC 11d]. To address this problem, we decided to explore more computationally efficient representations and methods. In particular, we investigated the replacement of the structural system model with a physics-agnostic but numerically accurate surrogate model based on non-uniform rational B-splines (NURBs) [STE 15], as illustrated in Figure 8.9. We describe here the numerical methodology for determining the proposed NURB-based surrogate models.

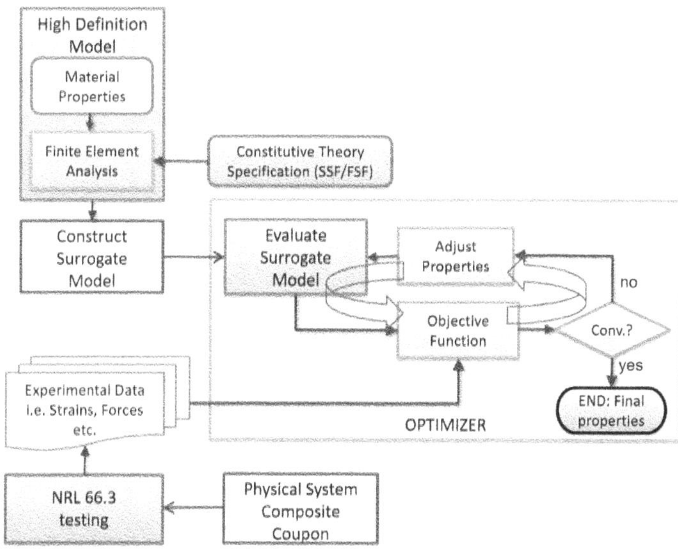

Figure 8.9. *Design optimization data flow for inverse characterization via utilization of a surrogate model*

In general, surrogate models are computationally efficient, but physics-agnostic representations are more expensive physics-specific models generated by sampling the underlying model and subsequently fitting some functional forms. A plethora of choices for surrogate models exists for our purposes [STE 15, TUR 05]. We used NURB-based surrogates due to the generality and computational efficiency of their formulation, as well as their established ability to represent nonlinear responses. For a particular geometry, loadings and common measurement locations and approximate solutions are sought on a system of the form:

$$\mathbf{t}^{\varepsilon}(\mathbf{x}_t) = \mathbf{c}^{\varepsilon}(\mathbf{a}, \mathbf{x}_c), \qquad [8.19]$$

where $\mathbf{t}^{\varepsilon}$ is a vector field function that collects experimentally observed strain tensor components and $\mathbf{c}^{\varepsilon}$ is a system model vector function of strain tensor components calculated by a proper method such as finite-element analysis. By introducing a surrogate model, $\mathbf{m}^{\varepsilon}$ of the original strain response model $\mathbf{c}^{\varepsilon}$ that involves the original constitutive model can be determined by equation [8.19] as:

$$\mathbf{m}^{\varepsilon}\left(\bar{\mathbf{a}}, \mathbf{x}\right) \approxeq \mathbf{t}^{\varepsilon}\left(\mathbf{x}\right) \qquad [8.20]$$

8.6.1. *Definition and construction of the surrogate model*

FEA-based system models evaluate the state variables onto a set of nodes inside the domain and on its surface. Thus, we assume that our sampling occurs on a discrete number of such points represented as vectors $\mathbf{x_i}, i = 1 \ldots n_m$. Consequently, our data consist of a set of n_m individual surrogate models, $\mathbf{m}_i^{\varepsilon}$, each of which approximates $\mathbf{c}^{\varepsilon}$ at a single point i (on the specimen surface S_o) represented by a vector $\mathbf{x}_i$:

$$\mathbf{m}_i^{\varepsilon}\left(\bar{\mathbf{a}}\right) = \mathbf{m}^{\varepsilon}\left(\bar{\mathbf{a}}, \mathbf{x}_i\right), i = 1 \ldots n_m \qquad [8.21]$$

The purpose of each of these surrogates is to generate a vector defined by its components as measurable scalars defined at various points of the observable part of the material specimen that depend on the constitutive parameters. We construct these surrogates in a manner that is based on the measured surface strain tensor components:

$$\mathbf{m}_i^{\varepsilon}\left(\bar{\mathbf{a}}\right) = \left\{{}_m^i\varepsilon_{xx}, {}_m^i\varepsilon_{yy}, {}_m^i\varepsilon_{xy}\right\}^T, i = 1 \ldots n_m \qquad [8.22]$$

Similarly, we define a set of experimental observations at corresponding locations, where the individual observations $\mathbf{t}_i$ have the same form as $\mathbf{m}_i$:

$$\mathbf{t}_i = \mathbf{t}\left(\mathbf{x}_i\right) = \left\{{}_t^i\varepsilon_{xx}, {}_t^i\varepsilon_{yy}, {}_t^i\varepsilon_{xy}\right\}^T, i = 1 \ldots n_m. \qquad [8.23]$$

For each of the nodes $\mathbf{x}_i \in S_o$, the system model $\mathbf{c}$ is sampled for l combinations of the constitutive parameters $\mathbf{a}_q, q = 1 \ldots l$ on a uniform lattice on the vector space of $\mathbf{a}$. This sampling gives a set of vector-valued elements $\mathbf{s}_j\left(\mathbf{a}_k\right)$ that correspond to the response of $\mathbf{c}^{\varepsilon}$ at a particular location on S_o.

We use a direct method to produce an exact interpolation of the acquired data to fit the NURB surrogates. In the case where $\mathbf{a}_q \in \mathbb{R}^1$

(hence, $\mathbf{a}_q = a_q$), this can be formulated as the solution to a set of simultaneous linear equations:

$$\mathbf{s}_j(a_q) = \mathbf{m}_j(u_q) = \mathbf{p}_j(u_q) = \sum_{i=1}^{n_c} {}^j\mathbf{b}_i N_{i,k}(u_q), \quad j = 1 \ldots n_m \text{ and } q = 1 \ldots l \qquad [8.24]$$

where $N_{i,k}(u_q)$ are the NURB basis functions and ${}^j\mathbf{b}_i$ are the unknown control point locations for the single variable case. The jth surrogate model $\mathbf{m}_j$ is defined as equal in value to the corresponding data set $\mathbf{s}_j$ at each of the parametric coordinate values of u_q. The values of u_q are normalized:

$$u_q = (a_q - a_{min}) / (a_{max} - a_{min}) \qquad [8.25]$$

The system of equations defined in equation [8.24] is extended to the multivariate case via the tensor product of NURB basis functions. Once the linear system has been formed, it may be solved using any technique. At the conclusion of the fitting process, $\mathbf{m}_i^\varepsilon$ are expressed as a set of interpolating functions:

$$\{\mathbf{m}_1^\varepsilon(\bar{\mathrm{a}}), \mathbf{m}_2^\varepsilon(\bar{\mathrm{a}}), \ldots, \mathbf{m}_{n_m}^\varepsilon(\bar{\mathrm{a}})\} = \{\mathbf{p}_1(\bar{\mathrm{a}}), \mathbf{p}_2(\bar{\mathrm{a}}), \ldots, \mathbf{p}_{n_m}(\bar{\mathrm{a}})\} \qquad [8.26]$$

8.6.2. *Characterization by optimization*

As indicated earlier, the characterization process can be formed as an inverse problem that seeks to minimize the difference between the experimentally observed and computationally calculated strain tensor components on the surface of a specimen. This is expressed in a discrete form as:

$$\begin{aligned} \text{min.} \quad & f(\bar{\mathbf{a}}) = \sum_{i=1}^{n_m} \|\mathbf{p}_i(\bar{\mathbf{a}}) - \mathbf{t}_i\| \\ \text{s.t.} \quad & a_{min_j} \leq \bar{a}_j \leq a_{max_j} \quad \forall \quad j \in 1 \ldots n_a \end{aligned} \qquad [8.27]$$

where n_a is the number of constitutive parameters (i.e. $\bar{\mathbf{a}} \in \mathbb{R}^{n_a}$). Figure 8.10 demonstrates this algorithm in a visual fashion.

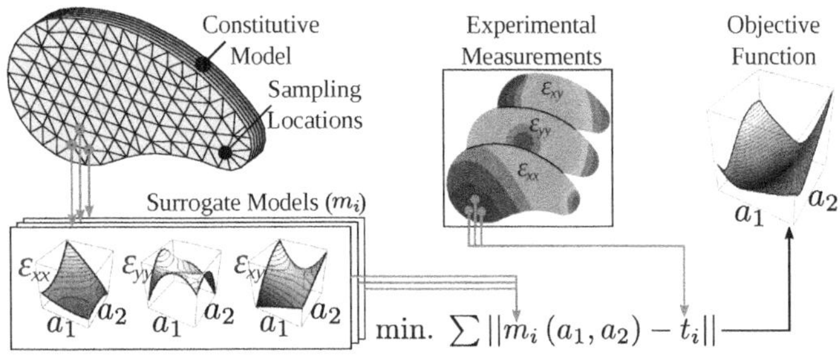

Figure 8.10. *Visual representation of the characterization algorithm using NURBs. For a color version of the figure, see www.iste.co.uk/brancherie/microstructure.zip*

In practice, any reasonably powerful nonlinear optimization routine can be employed. In this study, we have elected to use either simulated annealing [KIR 84] or the Nelder–Mead simplex algorithm [LAG 98], which do not require derivative information.

The presence of experimental noise complicates efforts to analyze the properties of f. In the formulation of equation [8.30], we now assume that the noise terms can be aggregated into a single error term ϵ, which will be considered to represent a random Gaussian noise with a mean of $\boldsymbol{\mu}_\epsilon = \mathbf{0}$ and variance of $\boldsymbol{s}_\epsilon$. The law of large numbers implies that a sufficiently large n_m will reduce the effect of experimental noise. The surrogate model formulation of the characterization problem can be used to combine several experimental *frames* of data, each corresponding to the same geometry, sampling locations and underlying constitutive model, but a different loading:

$$\text{min. } f(\bar{\mathbf{a}}) = \sum_{i=1}^{n_m} \left(w_1 ||\mathbf{m}^{\varepsilon}_{1,i}(\bar{\mathbf{a}}) - \mathbf{t}_{1,i}|| + w_2 ||\mathbf{m}^{\varepsilon}_{2,i}(\bar{\mathbf{a}}) - \mathbf{t}_{2,i}|| + \ldots \right)$$

$$\text{s.t. } a_{min_j} \leq \bar{a}_j \leq a_{max_j} \quad \forall \quad j \in 1 \ldots n_a \qquad [8.28]$$

where w_n is a weight parameter associated with the nth load case. This approach reaps additional benefit by making f more sensitive to the values of *all* of the constitutive parameters and hence avoids an ill-posed formulation [ILI 10].

	E_1(GPa)	E_2(GPa)	ν_{12}	ν_{23}	G_{12}(GPa)
Recovered	148.5	10.8	0.277	0.280	9.86
Reference	147.0	10.8	0.270	0.280	7.0

Table 8.2. *Recovery of constitutive properties from physical test data. The reference values for the parameters are taken from [DAN 05], in engineering form*

8.6.3. *Validation with physical experiments*

An experimental campaign was carried out using the automated test system NRL66.3 [MIC 11d] to characterize a series of AS4/3501-6 specimens. From this test campaign, 20 frames of data were chosen, one per specimen. The frames were taken from the load-path history of each experiment at a point where linear-orthotropic responses were dominant. The constitutive model used for characterization was implemented in the ANSYS FEA code. The results achieved from this process are shown in Table 6.

These results indicate that the surrogate model-based characterization can indeed be used to recover the anisotropic elastic properties of a composite from actual experimental data under real conditions. For this validation exercise, surrogate model construction required 1,015,600 s, and solving the optimization problem required 120 s. In a previous work [MIC 11d], we found that the optimization required 104,400 s. The break-even point, where the surrogate model approach yields a net improvement in computing time, thus requires only 10 experiments. Given that the campaigns associated with [MIC 11d] tested more than a thousand specimens, this is strong evidence that the surrogate-enabled methodology has practical value.

8.7. Multi-scale inversion

In order to achieve the inverse characterization of the composite material in multiple scales, we begin by modeling a representative volume element (RVE) of the composite structure at the micro-scale, aiming to estimate the average properties of the heterogeneous media in the meso-scale (lamina). Consequently, we have to first face the forward problem of calculating the homogenized properties of the given composite material, which, for the case of this chapter, we assume is composed of parallel, cylindrical fibers, embedded in a polymer matrix and exhibiting a periodic microstructure in either a square or hexagonal array.

Subsequently, the properties estimated by the aforementioned homogenization process are considered as the properties of the lamina at the meso-scale, which in turn forms an unsymmetric, balanced, angle-plied laminate at the macro-scale. Synthetic experiments performed on the laminate include six-axis mechanical loading, and they provide information regarding the developing strain fields. Solving the inverse structural problem involving the estimation of the composite material properties of the constituents at the micro-scale, by considering the homogenized response of the lamina in the meso-scale and the laminate in the macro-scale using experimental data collected at the macro-scale, through the use of appropriate optimization algorithms, enables the multi-scale inverse characterization of the composite material.

8.7.1. *Forward problem: mathematical homogenization*

Here, we focus on the three-dimensional modeling of the composite microstructure and utilize a micromechanics model that allows for the estimation of the full set of the elastic properties using a single model, as opposed to a collection of models accompanied by different assumptions in order to estimate the full set of the properties needed.

The problem of determining the material properties of heterogeneous materials, such as composites, has long been studied [ESH 57, HIL 65, BUD 65] with many analytical techniques of homogenization built upon the equivalent eigenstrain method [ESH 57], which investigates the problem of an ellipsoidal inclusion embedded in an infinite elastic medium.

The analytical methods mentioned so far provide approximate estimates of the average elastic properties of the macroscopically isotropic and sometimes anisotropic heterogeneous material. However, in order to comply with the transversely isotropic behavior that most composites exhibit due to the random distribution of fibers in the cross-section, an appropriate averaging procedure is used to provide a stiffness tensor with only five independent constants.

The current implementation involves only numerical homogenization [BAR 08] using finite-element analysis, in order to work toward the inverse problem of composite characterization at the micro-scale of constituents (fiber and matrix) in a manner that follows the established methodologies, as described in [BAR 13, YUA 08]. The composite is assumed to be composed of parallel cylindrical transversely isotropic fibers embedded in a polymer isotropic matrix. The microstructure is considered to be periodic and in either a square or a hexagonal arrangement, as shown in Figure 8.11. In order to estimate the effective elastic properties at the meso-scale, the composite is represented at the micro-scale by a 3D representative volume element (RVE), which exemplifies the arrangement of the constituents and is subjected to periodic boundary conditions [BAR 08].

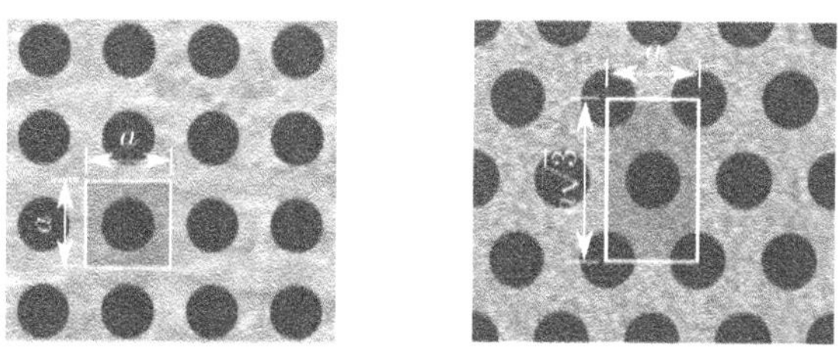

(a) Square arrangement (b) Hex arrangement

Figure 8.11. *Typical arrangements assumed in the homogenization of polymer matrix fiber-reinforced composite systems. Homogenization volume elements are shown in transparent red. For a color version of the figure, see www.iste.co.uk/brancherie/microstructure.zip*

Each RVE is subjected to an average strain $\bar{\epsilon}$ [LUC 98] and each of the six components of strain are applied by enforcing appropriate displacement boundary conditions through constraint equations at the faces, edges and vertices of the RVE [BAR 13]. The volume average of the strain ($\bar{\epsilon}_{ij}$) is equal to the applied strain (ϵ_{ij}^{app}):

$$\bar{\epsilon}_{ij} = \frac{1}{V}\int_V \epsilon_{ij} dV = \epsilon_{ij}^{app}. \qquad [8.29]$$

For the homogeneous composite, the average stress is calculated by (in Voigt notation):

$$\bar{\sigma}_\alpha = C_{\alpha\beta}\bar{\epsilon}_\beta. \qquad [8.30]$$

Consequently, the components of the Hooke tensor $C_{\alpha\beta}$ of the homogenized material can be determined by:

$$C_{\alpha\beta} = \bar{\sigma}_\alpha = \frac{1}{V}\int_V \sigma_\alpha (x, y, z)\, dV \text{ , with } \epsilon_\beta^{app} = 1, \qquad [8.31]$$

where x, y, z are the spatial coordinates. A detailed description of the process alongside the necessary constraint equations that satisfy periodicity for each loading case can be found in [BAR 13].

The components of C_{ij} for the case of transverse isotropy are given by [DAN 05, BAR 08, KOM 15].

8.7.2. *Inverse problem*

As indicated earlier, like in the single-scale characterization framework [MIC 12b, MIC 12a, STE 13], the inverse problem takes the form of an optimization problem. The main difference is the fact that the forward evaluation is applied through a composed form of the models through the various scales such that the use of the experimental measurements at the macro-scale is permissible.

The constituents of the micro-scale are considered to be the matrix, the fiber and the interface between the matrix and the fiber. We assume isotropic material properties of the matrix and the interface and transversely isotropic material properties for the fiber. The properties in consideration are then defined as: E_m, ν_m, Young's modulus and

Poisson's ratio of the matrix, respectively; E_i, ν_i, Young's modulus and Poisson's ratio of the interface, respectively, and E_{xf}, E_{yf}, G_{xyf}, ν_{xyf}, ν_{yzf}, the five independent engineering properties of the fiber. The set of properties is denoted as:

$$P = \{E_m, \nu_m, E_i, \nu_i, E_{xf}, E_{yf}, G_{xyf}, \nu_{xyf}, \nu_{yzf}\}. \qquad [8.32]$$

Because the aim of this study is to investigate the feasibility of identifying any subset of P: $U \subset P$ using an appropriate inversion scheme, we can now define the set difference $K = P \setminus U$ as the set of known properties before characterization. When needed, the values of the properties in the set U can be identified through simpler experiments, for example, a tension experiment to identify the matrix E_m and ν_m or the fiber axial Young's modulus, E_{xyf}.

It is anticipated that elements of P that cannot be known *a priori* will always exist. In the specific case of an RVE with an interface layer between the fiber and the matrix, these are the properties of the interface (E_i, ν_i) that manifest only after the composite layers have been manufactured. In addition, sometimes simple elastic material constants like ν_{yzf} cannot be determined experimentally with high confidence.

In the form described herein and with the aid of the definition of the forward problem, a block diagram of the multi-scale inverse characterization process is presented in the diagram shown in Figure 8.12. The entire process is inside an optimization loop that takes the homogenized parameters and performs yet another forward evaluation of the meso- and macro-scale models. This process is performed for each load case that full-field experimental data exist for. The strain results of the multi-scale forward analysis are then incorporated into an additive operation of objective functions. These objective functions are selected to be of the form:

$$f_k = \frac{1}{3N} \sum_{l=1}^{N} \sum_{i=x,y} \sum_{j=i,y} \left| \left[\epsilon_{ij}^{exp_k}\right]_l - \left[\epsilon_{ij}^{fea_k}\right]_l \right|, \qquad [8.33]$$

where $|\cdot|$ denotes the absolute value, $\epsilon_{ij}^{exp_k}$ are the surface strains measured at pre-determined locations via any full-field measurement method in general [SEV 81, BRU 89, SIR 91, AND 05] and with the MRG or DSI methods in particular [ILI 08, MIC 09a, ILI 12, MIC 11e, ILI 11, ILI 13] for the load case k, $\epsilon_{ij}^{fea_k}$ are the surface strains calculated from the forward finite-element analysis of the load case k, l is the id number of the location of the strain measurement and N is the total number of strain measurement locations. An aggregate objective function $f = 1/K \sum_{k=1}^{K} f_k$ is used to collect the values of each of the objective functions evaluated for the specific load cases.

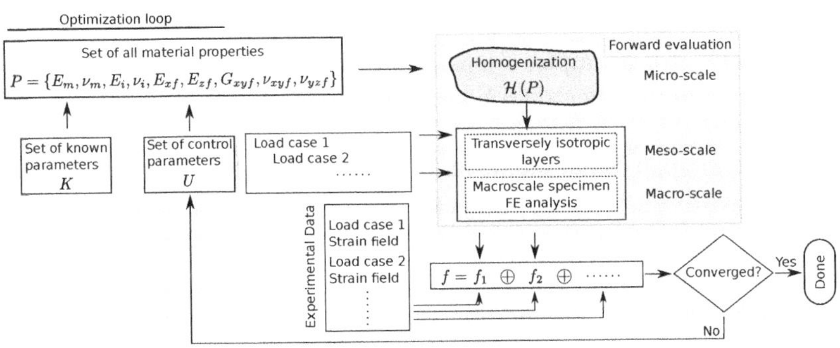

Figure 8.12. *Block diagram of the multi-scale inverse characterization process.*

8.8. Computational framework and synthetic experiments

The inverse methodology described in the previous section was implemented using MATLAB®[1] as the control language and ANSYS®Mechanical™[2] as the forward multi-scale evaluator. The connection between the two was achieved through an appropriate serial approach using file-system-based exchange of information.

1 MATLAB® is a registered trademark of The MathWorks, Inc.

2 ANSYS is a registered trademark of ANSYS, Inc.

Many optimization algorithms were explored as described in [KOM 15]. The one selected was a nonlinear derivative-free global optimization algorithm with proper modifications to constrain the search space of the optimization parameters [KOM 15].

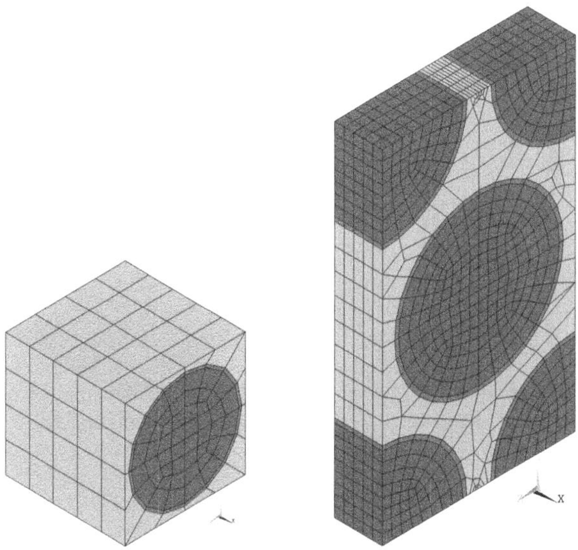

(a) Square (b) Hex

Figure 8.13. *Meshed unit cell geometries for a fiber-to-cell volume ratio of 0.63. Cyan: matrix; purple: fiber; red: interface layer. For a color version of the figure, see www.iste.co.uk/brancherie/microstructure.zip*

Figure 8.13 shows both the square and the hex cells utilized when an interface layer is present between the resin and the fiber.

The macro-scale geometry is presented in Figure 8.14(a). It has been discretized with hexahedral elements (SOLID185). A closeup view at the area of the hole is depicted in Figure 8.14(b). As it can be seen, the composite material layers are modeled with a single element through their thickness.

In order to investigate the capabilities of the presented multi-scale inverse method for characterizing a composite material at the constituents level by using data at the laminate level, we evaluated how accurately the elastic properties of the isotropic matrix and those of the transversely isotropic fiber can be estimated. This estimation was based on comparisons made relative to the synthetic data that were generated by running the forward problem with known constituent properties.

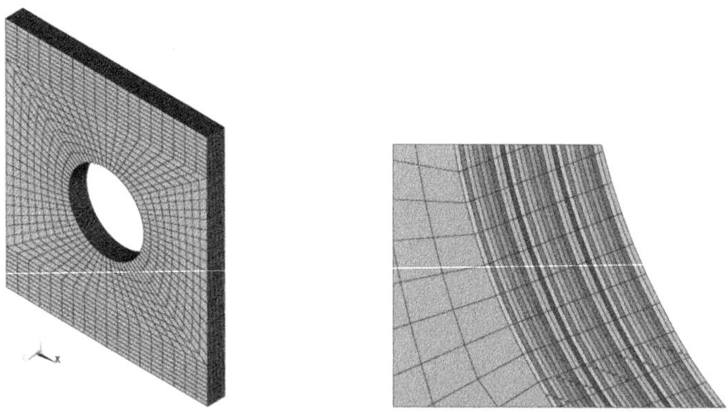

(a) General view of the meshed macro-scale specimen

(b) Close up showing the meso-scale layer configuration

Figure 8.14. *Meshed specimen geometry with an open hole. For a color version of the figure, see www.iste.co.uk/brancherie/microstructure.zip*

All 128 possible combinations of properties appearing in equation [8.32] were examined assuming that the interface properties (E_i, ν_i) are always unknown. With regard to the macro-scale model, the laminate specimen dimensions were chosen to be: height, 45 mm; width, 40 mm; thickness, 3.5 mm and hole diameter, 15 mm. For all the synthetic experiments, a single loading case was selected, defined in the full loading space of a 6 degrees-of-freedom loading vector acting upon the laminate at macro-scale [ILI 10]. The values of the components of this vector for the three force components and the three components of moment were: $F_x = 2241\ N$, $F_y = 4310\ N$, $F_z = 517\ N$, $M_x = 3.45\ Nm$, $M_y = 6.9\ Nm$ and $M_z = 69\ Nm$. The magnitude of these loads was chosen to ensure that the developed corresponding levels of strain respect the linear elastic limits of the material model.

	E_m	ν_m	E_{xf}	E_{yf}	
Value	$4.3GPa$	0.35	$235GPa$	$15GPa$	
min	$0.5GPa$	0.0	$10GPa$	$1GPa$	
max	$15.0GPa$	0.45	$600GPa$	$60GPa$	
	ν_{xyf}	ν_{yzf}	G_{xyf}	E_i	ν_i
Value	0.2	0.07	$27GPa$	$2.3GPa$	0.25
min	0.0	0.0	$1GPa$	$0.1GPa$	0.0
max	0.45	0.45	$75GPa$	$5GPa$	0.45

Table 8.3. *Constituents material properties and optimization bounds*

The composite material used to generate the synthetic results is the AS4/3501-6 carbon epoxy, with values of elastic properties of the fiber, the matrix and the lamina as they appear in [DAN 05] and also presented in Table 8.3, together with assumed properties for the interface layer. The error for each material property reported in the relevant tables were calculated using the formula:

$$e = \left| \frac{P_{inverse} - P_{nominal}}{P_{nominal}} \right| \times 100\,[\%] \qquad [8.34]$$

where $P_{inverse}$ is the value of the property identified by the multi-scale inverse characterization and $P_{nominal}$ is the value used to generate the synthetic data.

Starting by deploying the methodology described above and utilizing the associated computational framework for only one parameter and working our way up to the total number of unknown parameters and their possible combinations, we were able to obtain a maximum number of unknown parameters which can be concurrently defined with high accuracy. We have assumed an isotropic interface zone under the condition that the matrix and fiber properties are known.

The parameter identification was successful, even for values of the interface properties as low as 0.5 % of the matrix's elastic properties. In the absence of noise in the synthetically generated data, the estimation of the interface properties can be achieved with very low error.

ID	E_m	ν_m	E_{xf}	E_{yf}	ν_{xyf}	ν_{yzf}	G_{xyf}	E_i	ν_i	Evaluations	f $(pStrain)$
sia	-	-	-	-	-	-	-	$< 0.1\%$	$< 0.1\%$	124	2.3
sib	-	-	$< 0.1\%$	-	-	$< 0.1\%$	$< 0.1\%$	$< 0.1\%$	$< 0.1\%$	1346	2.1
sic	-	0.13%	-	$< 0.1\%$	$< 0.1\%$	-	$< 0.1\%$	$< 0.1\%$	0.8%	3231	1.7
sid	0.3%	4.2%	-	-	-	1.2%	4.6%	0.65%	35.0%	2605	52.8

Table 8.4. *Multi-scale inverse characterization for selected cases of the square volume element with interface layer synthetic experiments. Case IDs represent different choices of unknown parameters*

In Table 8.4, we present some interesting inversion results for the case of the square with interface RVE. In this table, we observe that the interface properties can be found with high accuracy if the properties of the other constituents are known, which may indeed reflect a realistic situation when constituent properties can be determined experimentally. The same is true for other cases like the *sib* case. As the number of unknowns increases, the results get worse. The analysis of all 128 cases revealed that no case with more than six unknowns was inverted efficiently.

ID	E_m	ν_m	E_{xf}	E_{yf}	ν_{xyf}	ν_{yzf}	G_{xyf}	E_i	ν_i	Evaluations	f $(pStrain)$
hia	-	-	-	-	-	-	-	$< 0.1\%$	$< 0.1\%$	139	2.0
hib	-	-	$< 1.3\%$	-	-	19%	10.4%	4.2%	52%	485	509910
hic	-	1.2%	-	$< 0.1\%$	1.6%	-	0.6%	0.1%	6.3%	1430	2.1
hid	2.2%	7.1%	-	-	-	8.1%	4.6%	2.4%	35.4%	1485	347679

Table 8.5. *Multi-scale inverse characterization for selected cases of the hex volume element with interface-layer synthetic experiments*

The results of the hex RVE with interface layer are presented in Table 8.5. It is observed that in this case, the interface properties are identified when they are the only unknowns. For other cases, the properties cannot be identified with adequate accuracy, besides the *hic* case. It should be noted that this is only a subset of interesting results and that other cases have been solved with high accuracy.

8.9. Conclusions and plans

In this chapter, we presented a methodological framework along with its computational implementation which can be used to identify both macro- and lower-scale properties of composite materials from experiments conducted on higher scales. The macro-scale applications included both finite and infinitesimal strain formulations. In addition, its extension to use NURB-based surrogate models was presented and validated with respect to synthetic and actual experiments for the macro-scale constitutive characterization of composites.

By replacing the expensive constitutive model with an efficient surrogate in the optimization loop which drives the characterization process, material properties can be recovered in seconds, as opposed to the previously required hours or days. We have also shown that the expense of building the surrogate models is a one-time expense and that a single surrogate model can be used to characterize several material systems. In addition, we have demonstrated that the NURB-based surrogates offer a convenient method for data fusion and allow the results from many experiments to be combined in order to achieve characterization. Finally, we have demonstrated that this algorithm meets our expectations of accuracy in the presence of experimental noise and can also properly characterize physical specimens using data acquired during a robotic test campaign. We believe that their strength prompts further investigation into techniques for material characterization enabled by surrogate models.

We demonstrated the feasibility of identifying mechanical properties of constituents of composite materials at the micro-scale from experiments conducted at the macro-scale. The most interesting case is the identification of the properties of constituents that manifest only after the medium has been manufactured and could not be known *a priori* or through experimentation. We investigated the inversion methodology on the identification of properties of the interface layer that develops between the matrix and the fibers of a polymer matrix fiber-reinforced composite system. Both Young's modulus and Poisson's ratio were identified for the synthetic experiments conducted. Furthermore, it was also shown that it is possible to identify properties of other constituents as long as some of them are known.

In the future, we will conduct a sensitivity analysis with respect to the constituent material properties as well as investigate additional geometrical parameters as measures of damage mechanism evolution and incorporate the uncertainty of realistic (random) fiber distributions to eliminate the assumptions of periodic fiber arrays. We will also apply the proposed methodology by utilizing various global optimization methodologies in an attempt to establish a more detailed and efficient implementation of the inversion problem.

8.10. Acknowledgments

The authors acknowledge support for this study from the Office of Naval Research (ONR) through the Naval Research Laboratory's core funding. Constructive suggestions by Dr. Nathan Bechle, Forest Products Laboratory in Madison Wisconsin, are greatly appreciated.

8.11. Bibliography

[AND 05] ANDRIANOPOULOS N.P., ILIOPOULOS A., "Strain measurements by a hybrid experimental-numerical method using a mesh-free field function", *Honorary Volume for Professor P. S. Theocaris*, pp. 31–41, Armenian Academy of Sciences, Yerevan, 2005.

[AND 06] ANDRIANOPOULOS N.P., ILIOPOULOS A., "Displacements measurement in irregularly bounded plates using mesh free methods", *Proc. 16th European Conference of Fracture*, Alexandroupolis, Greece, 3–7 July 2006.

[BAR 08] BARBERO E.J., *Finite Element Analysis of Composite Materials*, CRC Press, Taylor & Francis Group, 2008.

[BAR 13] BARBERO E.J., *Finite Element Analysis of Composite Materials Using ANSYS*, CRC Press, Taylor & Francis Group, 2013.

[BRU 89] BRUCK H., MCNEILL S., SUTTON M. *et al.*, "Digital image correlation using Newton–Raphson method of partial differential correction", *Experimental Mechanics*, vol. 29, pp. 261–267, 1989. doi: 10.1007/BF02321405.

[BRU 02] BRUNO L., FURGIUELE F.M., PAGNOTTA L. *et al.*, "A full-field approach for the elastic characterization of anisotropic materials", *Optics and Lasers in Engineering*, vol. 37, no. 4, pp. 417–431, 2002.

[BUD 65] BUDIANSKY B., "On the elastic moduli of some heterogeneous materials", *Journal of the Mechanics and Physics of Solids*, vol. 13, pp. 223–227, 1965.

[CAR 02] CARLSSON L., ADAMS D.F., PIPES R.B., *Experimental Characterization of Advanced Composite Materials, Third Edition*, 3rd edition, CRC Press, 2002.

[COO 07] COOREMAN S., LECOMPTE D., SOL H. *et al.*, "Elasto-plastic material parameter identification by inverse methods: calculation of the sensitivity matrix", *International Journal of Solids and Structures*, vol. 44, no. 13, pp. 4329–4341, 2007.

[DAN 05] DANIEL I.M., ISHAI O., *Engineering Mechanics of Composite Materials*, 2nd edition, Oxford University Press, New York, 2005.

[ESH 57] ESHELBY J.D., "The determination of the elastic field of an ellipsoidal inclusion, and related problems", *Proceedings of the Royal Society of London A: Mathematical, Physical and Engineering Sciences*, vol. 241, no. 1226, pp. 376–396, 1957.

[FU 16] FU Y., MICHOPOULOS J.G., SONG J.-H., "On investigating the thermomechanical properties of cross-linked epoxy via molecular dynamics analysis", *Nanoscale and Microscale Thermophysical Engineering*, vol. 21, no. 1, pp. 8–25, 2016.

[HIL 65] HILL R., "A self-consistent mechanics of composite materials", *Journal of the Mechanics and Physics of Solids*, vol. 13, no. 4, pp. 213–222, 1965.

[ILI 07] ILIOPOULOS A., Digital Image Processing on Experimental Mechanics and Applications on Metallic Sheets, PhD Thesis, National Technical University of Athens, 2007.

[ILI 08] ILIOPOULOS A., ANDRIANOPOULOS N.P., "An approach to analyse errors introduced in the random grid strain measurement method", *Strain*, vol. 46, no. 3, pp. 258–266, 2008.

[ILI 09a] ILIOPOULOS A., MICHOPOULOS J., "Sensitivity analysis of the mesh-free random grid method for measuring deformation fields on composites", *Proceedings of the 17th International Conference on Composite Materials, ICCM-17, Edinburgh*, The British Composites Society: Edinburgh International Convention Centre (EICC), Edinburgh, 27–31 July 2009.

[ILI 09b] ILIOPOULOS A., MICHOPOULOS J., "Effects of anisotropy on the performance sensitivity of the mesh-free random grid method for whole field strain measurement", *ASME Conference Proceedings*, vol. 2009, no. 48999, pp. 65–74, January 2009.

[ILI 10] ILIOPOULOS A., MICHOPOULOS J., "Loading subspace selection for multidimensional characterization tests via computational experiments", *Proceedings of the ASME 2010 International Design Engineering Technical Conferences & Computers and Information in Engineering Conference IDETC/CIE 2010*, no. DETC2010/CIE-28741, ASME, 2010.

[ILI 11] ILIOPOULOS A., MICHOPOULOS J., ORIFICI A.C. *et al.*, "Experimental validation of the 2D meshless random grid method", *ASME Conference Proceedings*, vol. 2011, no. 54792, pp. 513–520, 2011.

[ILI 12] ILIOPOULOS A., MICHOPOULOS J., ANDRIANOPOULOS N.P., "Performance analysis of the mesh-free random grid method for full-field synthetic strain measurements", *Strain*, vol. 48, no. 1, pp. 1–15, 2012.

[ILI 13] ILIOPOULOS A., MICHOPOULOS J., "Direct strain tensor approximation for full field strain measurement methods", *International Journal of Numerical Methods in Engineering*, vol. 95, pp. 313–330, 2013.

[ILI 14] ILIOPOULOS A., MICHOPOULOS J., "Meshless methods for full field displacement and strain measurement", in MICHOPOULOS J.G., ROSEN D.W., PAREDIS C.J.J. *et al.* (eds), *Advances in Computers and Information in Engineering Research*, vol. 1, pp. 96–115, ASME, New York, 2014.

[ILI 16] ILIOPOULOS A., STEUBEN J.C., MICHOPOULOS J., "Determination of anisotropic mechanical properties of G-10 composite via Direct Strain Imaging", *Polymer Testing*, vol. 50, pp. 64–72, 2016.

[JU 98] JU J.W., CHABOCHE J., VOYIADJIS G.Z., *Damage Mechanics in Engineering Materials*, 1st edition, Elsevier Science, January 1998.

[KAJ 04] KAJBERG J., LINDKVIST G., "Characterisation of materials subjected to large strains by inverse modelling based on in-plane displacement fields", *International Journal of Solids and Structures*, vol. 41, no. 13, pp. 3439–3459, 2004.

[KAL 00] KALISKE M., "A formulation of elasticity and viscoelasticiy for fibre reinforced material at small and finite strains", *Computer Methods in Applied Mechanics and Engineering*, vol. 185, pp. 225–243, 2000.

[KIR 84] KIRKPATRICK S., "Optimization by simulated annealing: Quantitative studies", *Journal of statistical physics*, vol. 34, no. 5, pp. 975–986, 1984.

[KOM 15] KOMNINELLI F., ILIOPOULOS A., MICHOPOULOS J., "Towards identification of lower scale composite material properties from higher scale experimental data via inverse analysis of coupled multiscale models", *Proceedings of the ASME 2015 International Design Engineering Technical Conferences & Computers and Information in Engineering Conference*, Boston, 2–5 August 2015.

[LAG 98] LAGARIAS J.C., REEDS J.A., WRIGHT M.H. *et al.*, "Convergence properties of the Nelder–Mead simplex method in low dimensions", *SIAM Journal on optimization*, vol. 9, no. 1, pp. 112–147, 1998.

[LIU 09] LIU G.-R., *Meshfree Methods: Moving Beyond the Finite Element Method*, Taylor & Francis, Abingdon, 2009.

[LUC 98] LUCIANO R., SACCO E., "Variational methods for the homogenization of periodic heterogeneous media", *European Journal of Mechanics – A/Solids*, vol. 17, no. 4, pp. 599–617, 1998.

[MAN 08] MAN H., FURUKAWA T., KELLERMANN D., "Implicit constitutive modeling based on the energy principles", *Proceedings of XXII ICTAM*, Adelaide, 25–59 August 2008.

[MAS 92] MAST P.W., NASH G.E., MICHOPOULOS J.G. *et al.*, Experimental determination of dissipated energy density as a measure of strain-induced damage in composites, Report no. Tec. Rpt. NRL/FR/6383–92-9369, Naval Research Laboratory, Washington, DC, 1992.

[MAS 95] MAST P., NASH G., MICHOPOULOS J. *et al.*, "Characterization of strain-induced damage in composites based on the dissipated energy density. Part I. Basic scheme and formulation", *Theoretical and Applied Fracture Mechanics*, vol. 22, no. 2, pp. 71–96, 1995.

[MEU 98] MEUWISSEN M.H.H., OOMENS C.W.J., BAAIJENS F.P.T. *et al.*, "Determination of the elasto-plastic properties of aluminium using a mixed numerical-experimental method", *Journal of Materials Processing Technology*, vol. 75, nos. 1–3, pp. 204–211, 1998.

[MIC 03] MICHOPOULOS J., "Computational and mechatronic automation of multiphysics research for structural and material systems", in *Recent Advances in Composite Materials*, Kluwer Academic Press, 2003.

[MIC 04] MICHOPOULOS J., "Mechatronically automated characterization of material constitutive response", *Proceedings of the 6th World Congress on Computational Mechanics (WCCM-VI)*, Tsinghua University Press and Springer, pp. 486–491, Beijing, China, 5–10 September 2004.

[MIC 08] MICHOPOULOS J., HERMANSON J.C., FURUKAWA T., "Towards the robotic characterization of the constitutive response of composite materials", *Composite Structures*, vol. 86, nos. 1–3, pp. 154–164, 2008.

[MIC 09a] MICHOPOULOS J., ILIOPOULOS A., "A computational workbench for remote full field 2D displacement and strain measurements", *ASME Conference Proceedings*, vol. 2009, no. 48999, pp. 55–63, 2009.

[MIC 09b] MICHOPOULOS J., ILIOPOULOS A., FURUKAWA T., "Accuracy of inverse composite laminate characterization via the mesh free random grid method", *ASME Conference Proceedings*, vol. 2009, no. 48999, pp. 367–374, 2009.

[MIC 09c] MICHOPOULOS J., FURUKAWA T., LAMBRAKOS S., "Data-driven characterization of composites based on virtual deterministic and noisy multiaxial data", *2008 ASME International Design Engineering Technical Conferences and Computers and Information in Engineering Conference, DETC 2008*, vol. 3, pp. 1095–1106, New York, 2009.

[MIC 10a] MICHOPOULOS J., HERMANSON J.C., ILIOPOULOS A., "Towards a recursive hexapod for the multidimensional mechanical testing of composites", *ASME Conference Proceedings*, vol. 2010, no. 44113, pp. 91–99, ASME, 2010.

[MIC 10b] MICHOPOULOS J., LAMBRAKOS S.G., ILIOPOULOS A., "On a data and requirements driven multi-scale framework linking performance to materials", *ASME Conference Proceedings*, vol. 2010, no. 44113, pp. 197–210, ASME, 2010.

[MIC 11a] MICHOPOULOS J., HERMANSON J.C., ILIOPOULOS A. *et al.*, "Overview of constitutive response characterization for composite materials via data-driven design optimization", *Proceedings of the ASME 2011 International Design Engineering Technical Conferences & Computers and Information in Engineering Conference IDETC/CIE*, Washington, DC, 29–31 August, 2011.

[MIC 11b] MICHOPOULOS J., ILIOPOULOS A., "A computational workbench for remote full field 3D displacement and strain measurements", *Proceedings of the ASME 2011 International Design Engineering Technical Conferences & Computers and Information in Engineering Conference IDETC/CIE*, Washington, DC, 29–31 August, 2011.

[MIC 11c] MICHOPOULOS J., HERMANSON J.C., ILIOPOULOS A. *et al.*, "Data-driven design optimization for composite material characterization", *Journal of Computing and Information Science in Engineering*, vol. 11, no. 2, 2011.

[MIC 11d] MICHOPOULOS J., HERMANSON J.C., ILIOPOULOS A. *et al.*, "Data-driven design optimization for composite material characterization", *Journal of Computing and Information Science in Engineering*, vol. 11, no. 2, p. 021009, 2011.

[MIC 11e] MICHOPOULOS J., ILIOPOULOS A., "A computational workbench for remote full field 3D displacement and strain measurements", *ASME Conference Proceedings*, vol. 2011, no. 54792, pp. 489–498, ASME, 2011.

[MIC 12a] MICHOPOULOS J., HERMANSON J.C., ILIOPOULOS A., "Robotic and multiaxial testing for the constitutive characterization of composites", *Materials Challenges in Alternative & Renewable Energy*, Clearwater, FL, February 26–March 1, 2012.

[MIC 12b] MICHOPOULOS J., HERMANSON J.C., ILIOPOULOS A., "First industrial strength multi-axial robotic testing campaign for composite material characterization", *2012 ASME International Design Engineering Technical Conferences & Computers and Information in Engineering Conference*, Chicago, IL, 12–15 August, 2012.

[MIC 14] MICHOPOULOS J., HERMANSON J.C., ILIOPOULOS A., "Advances on the constitutive characterization of composites via multiaxial robotic testing and design optimization", in MICHOPOULOS J.G., ROSEN D.W., PAREDIS C.J.J. *et al.* (eds), *Advances in Computers and Information in Engineering Research*, vol. 1, ASME, New York, 2014.

[MOL 05] MOLIMARD J., LE RICHE R., VAUTRIN A. *et al.*, "Identification of the four orthotropic plate stiffnesses using a single open-hole tensile test", *Experimental Mechanics*, vol. 45, pp. 404–411, 2005. doi: 10.1007/BF02427987.

[PIE 07] PIERRON F., VERT G., BURGUETE R. *et al.*, "Identification of the orthotropic elastic stiffnesses of composites with the virtual fields method: sensitivity study and experimental validation", *Strain*, vol. 43, no. 3, pp. 250–259, 2007.

[RYA 90] RYAN K.F., "Dynamic response of graphite/epoxy plates subjected to impact loading", Thesis (M.S.), Massachusetts Institute of Technology, Dept. of Aeronautics and Astronautics, available at: http://dspace.mit.edu/handle/1721.1/42982, 1990.

[SCH 03] SCHMIDT T., TYSON J., GALANULIS K., "Full–field dynamic displacement and strain measurement using advanced 3d image correlation photogrammetry: part 1", *Experimental Techniques*, vol. 27, no. 3, pp. 47–50, 2003.

[SEV 81] SEVENHUIJSEN J., "Two simple methods for deformation demonstration and measurement", *Strain*, vol. 17, no. 1, pp. 20–24, 1981.

[SIR 91] SIRKIS J., LIM T., "Displacement and strain measurement with automated grid methods", *Experimental Mechanics*, vol. 31, pp. 382–388, 1991. doi: 10.1007/BF02325997.

[STE 13] STEUBEN J., MICHOPOULOS J., ILIOPOULOS A. *et al.*, "Inverse characterization of composite materials using surrogate models", *ASME 2013 International Design Engineering Technical Conferences & Computers and Information in Engineering Conference IDETC/CIE 2013*, no. DETC/CIE2013–12656, ASME Press, Portland, Order No.: I924DV, pp. 1–12, August 2013.

[STE 15] STEUBEN J., MICHOPOULOS J., ILIOPOULOS A., *et al.*, "Inverse characterization of composite materials via surrogate modeling", *Composite Structures*, vol. 132, pp. 694–708, 2015.

[TUR 05] TURNER C.J., CRAWFORD R.H., "Selecting an appropriate metamodel: the case for NURBs metamodels", *ASME 2005 International Design Engineering Technical Conferences and Computers and Information in Engineering Conference*, American Society of Mechanical Engineers, pp. 759–771, 2005.

[VAI 97] VAIDYA R., SUN C., "Fracture criterion for notched thin composite laminates", *AIAA Journal*, vol. 35, no. 2, pp. 311–316, 1997.

[WEG 00] WEGNER P.M., ADAMS D.F., Verification of the Combined Load Compression (CLC) Test Methods, Report no. DOT/FAA/AR-00/26, DOT/FAA, 2000.

[YUA 08] YUAN Z., FISH J., "Toward realization of computational homogenization in practice", *International Journal for Numerical Methods in Engineering*, vol. 73, no. 3, pp. 361–380, 2008.

9

New Trends in Computational Mechanics: Model Order Reduction, Manifold Learning and Data-Driven

At the moment, engineering sciences and technology is experiencing a data revolution. In the past, models were more abundant than data, the latter being too expensive to be collected and analyzed. However, the situation is how radically different. Data are much more abundant (and sometimes accurate) than existing models, and a new paradigm is emerging in engineering sciences and technology. In this chapter, we retrace some incipient applications on the basis of data within the framework of computational mechanics. The following three main topics are addressed in this chapter: (i) construction of solution manifolds and its use for interpolating new solutions on the manifold; (ii) construction of parametric solutions on the just extracted manifold and (iii) definition of behavior manifolds to perform data-driven simulation while avoiding the use of usual constitutive equations.

Chapter written by Jose Vicente AGUADO, Domenico BORZACCHIELLO, Elena LOPEZ, Emmanuelle ABISSET-CHAVANNE, David GONZALEZ, Elias CUETO and Francisco CHINESTA.

9.1. Introduction

9.1.1. *The big picture*

Advanced model order reduction techniques have allowed us to move from data to information, because in many systems, and despite the large amount of data, the hidden information was quite reduced, and it has been successfully extracted by applying many current state-of-the-art model reduction techniques (POD, PGD or RB among many other variants).

Before solving a given problem, the user must introduce the different involved parameters (e.g. material parameters and applied loads) as well as define the domain in which the problem is posed. However, this procedure has one main drawback: it rarely allows to progress in real time. In these circumstances, the real-time performance required in some applications is compromised. We could think that all these issues could be circumvented with the mere use of more powerful computers. Even if this is a possibility, it compromises the accessibility to the appropriate simulation resources of small and medium-sized industries. In order to democratize simulation, new solutions are required. A possible alternative consists of calculating offline (using all the needed computational resources and computing time) a parametric solution containing the solution of all possible scenarios, which is then particularized online using light computational facilities, such as deployed devices, tablets or even smartphones, for performing efficient simulation, optimization, inverse analysis, uncertainty propagation and simulation-based control, all under real-time constraints.

Even if we could think that for constructing the above-described parametric solutions it is enough to solve the model at hand for any possible choice of the parameters that it involves, it is clear that such a procedure rapidly fails because it involves a combinatorial explosion (e.g. 10 parameters each taking 10 possible values will involve 10 to the power of 10 possibilities, and 10 parameters taking 10 possible values remains too simplistic in applications of practical interest).

Recently, model order reduction has opened up new possibilities. First, proper orthogonal decompositions (POD) allows to extract the most significant characteristics of the solution, which can be then applied in solving models slightly different to the ones that served to define the reduced approximation basis, by simply projecting the searched solution onto the extracted reduced approximation basis [CHI 16].

Another family of model reduction techniques lies in the use of reduced bases constructed by combining a greedy algorithm and *a priori* error indicators. It involves some offline work, but then the reduced basis can be used online for solving different models with a perfect control of the solution accuracy because of the availability of error bounds. When the error is inadmissible, the reduced basis can be enriched by reusing the same greedy algorithm [CHI 16].

Finally, proper generalized decomposition methods are based on the use of separated representations [LAD 85, LAD 96, AMM 06, AMM 10]. Such separated representations are considered when solving at-hand partial differential equations by employing procedures based on the separation of variables, as will be described in the next section.

Advanced clustering techniques not only help engineers and analysts, but are also crucial in many areas where models, approximation bases, parameters and so forth are adapted depending on the local (in space and time senses) state of the system. They make defining hierarchical and goal-oriented modeling possible.

Machine and manifold learning also helps to extract the manifold in which the solutions of complex and coupled engineering problems are defined. Thus, uncorrelated parameters can be efficiently extracted from the collected data coming from numerical simulations, experiments or even from the data collected from adequate measurement devices. As soon as uncorrelated parameters are identified (constituting the information level), the solution of the problem can be predicted in new points of the parametric space, from adequate interpolations or otherwise, parametric solutions can be obtained within an adequate framework able to circumvent the curse of dimensionality (combinatorial explosion) for any value of the uncorrelated model parameters. Thus, the subtle circle is closed by linking data to information, information to knowledge and finally knowledge to real-time decision-making, opening unimaginable possibilities within the so-called DDDAS (dynamic data-driven application systems) that allows even model-free simulations.

All the above-mentioned techniques will be revisited in this chapter.

9.1.2. *The PGD at a glance*

Most of the existing model reduction techniques proceed by extracting a suitable reduced basis and then projecting onto it the problem solution. Thus, the reduced basis construction precedes its use in the solution procedure, and we must be careful concerning the suitability of a particular reduced basis when used for representing the solution of a particular problem. This issue disappears if the approximation basis is constructed and the problem is solved at the same time. Thus, each problem has its associated basis in which its solution is expressed. We could consider a few terms in its approximation, leading to a reduced representation, or all the terms needed for approximating the solution up to a certain accuracy level. The proper generalized decomposition (PGD) proceeds in this manner.

When calculating the transient solution of a generic problem $u(x, t)$, we usually consider a given basis of space functions $N_i(x)$,

$i = 1, \cdots, \mathcal{N}$, the so-called shape functions within the finite-element framework, and approximate the problem solution as:

$$u(x,t) \approx \sum_{i=1}^{\mathcal{N}} a_i(t) N_i(x), \qquad [9.1]$$

which implies a space-time-separated representation, where the time-dependent coefficients $a_i(t)$ are unknown at each time (when proceeding incrementally) and the space functions $N_i(x)$ are given *a priori*, for example, polynomial basis. POD and reduced bases methodologies consider a reduced basis $\phi_i(x)$ for approximating the solution instead of using the generic functions $N_i(x)$. The former are expected to be more suitable for approximating the problem at hand. Thus, it results:

$$u(x,t) \approx \sum_{i=1}^{R} b_i(t) \phi_i(x), \qquad [9.2]$$

where generally $R \ll \mathcal{N}$. Again, [9.2] is a space-time-separated representation, where the time-dependent coefficient must be calculated at each time during the incremental solution procedure.

Inspired from these results, we could consider the general space-time-separated representation:

$$u(x,t) \approx \sum_{i=1}^{N} X_i(x) T_i(t), \qquad [9.3]$$

where now neither the time-dependent functions $T_i(t)$ nor the space functions $X_i(x)$ are *a priori* known. Both will be computed simultaneously during problem-solving.

As soon as we postulate that the solution of a transient problem can be expressed in the separated form [9.3] whose approximation functions $X_i(x)$ and $T_i(t)$ will be determined during the problem solution, we take a step forward and assume that the solution of a multi-dimensional problem $u(x_1, \cdots, x_d)$ could be found in the separated form:

$$u(x_1, x_2, \cdots, x_d) \approx \sum_{i=1}^{N} X_i^1(x_1) X_i^2(x_1) \cdots X_i^d(x_d), \qquad [9.4]$$

and further more, expressing the 3D solution $u(x, y, z)$ as a finite sum decomposition involving lower-dimensional functions:

$$u(x, y, z) \approx \sum_{i=1}^{N} X_i(x) Y_i(y) Z_i(z), \qquad [9.5]$$

or:

$$u(x, y, z) \approx \sum_{i=1}^{N} X_i(x, y) Z_i(z), \qquad [9.6]$$

and the solution of a parametric problem $u(\mathbf{x}, t, p_1, \cdots, p_\wp)$ as:

$$u(\mathbf{x}, t, p_1, \cdots, p_\wp) \approx \sum_{i=1}^{N} X_i(\mathbf{x}) T_i(t) \prod_{k=1}^{\wp} P_i^k(p_k). \qquad [9.7]$$

The performances of all these separated representations are quite impressive in many cases.

For a review on such techniques and their applications in engineering sciences, the interested reader can refer to [CHI 10, CHI 11, CHI 13a, CHI 13b] and the numerous references therein concerning the space-time/space-frequency decomposition [BOU 97, LAD 99, AMM 07, AMM 11, BOU 13, RIO 13, BAR 14], space separation [BOG 12, VID 12, VID 13, BOG 14, BOR 15] and parametric solutions [CHI 13b, HEY 13, AMM 14, NER 15], allowing real-time simulations [CHI 13b, GON 14, GON 15], optimization [GHN 11, CHI 13b, CHI 14, BOR 17, AGU 17], simulation-based control [GHN 12, GON 12, CHI 13b, AGU 15], uncertainty propagation [NOU 08, NOU 09a, NOU 09b, NOU 10] or multi-scale upscaling [LAM 10, NER 10, CRE 13].

In all the applications referenced above, the choice of parameters was a simple matter: material or process parameters, boundary conditions and so on. However, in other applications, the extraction of uncorrelated model parameters is not an easy task, as is the case, for example when addressing shape parametrization or the description of microstructures. These issues will be addressed in the following sections.

9.2. Constructing slow manifolds

It is well known that microstructures or shapes do not allow simple reduced descriptions. The main question is not whether microstructures or shapes define slow manifolds or not, but whether they can be parametrized, that is, represented, or not, from slow manifolds. The same issue applies in visualization of high-dimensional data.

9.2.1. *From principal component analysis (PCA) to kernel principal component analysis (kPCA)*

Let us consider D observed variables defining the vector $\mathbf{X} \in \mathbb{R}^D$. These are commonly referred to in the MOR literature as *snapshots*. We assume that these variables are therefore not uncorrelated and, notably, that there exists a linear transformation $\mathbf{W}$ defining the vector $\mathbf{Y} \in \mathbb{R}^d$, where $d < D$ represents the unknown so-called *latent variables*, according to:

$$\mathbf{X} = \mathbf{W}\mathbf{Y}. \quad [9.8]$$

The transformation $\mathbf{W}$, $D \times d$, is assumed to verify the orthogonality condition $\mathbf{W}^T\mathbf{W} = \mathbf{I}_d$, where $\mathbf{I}_d$ represents the $d \times d$-identity matrix ($\mathbf{W}\mathbf{W}^T$ is not necessarily $\mathbf{I}_D$). The existence of such a transformation is precisely at the origin of PCA methods.

We assume the existence of M different snapshots $\mathbf{X}_1, \ldots, \mathbf{X}_M$ that can be stored in the columns of the $D \times M$ matrix $\mathbb{X}$. The associated $d \times M$ reduced matrix $\mathbb{Y}$ contains the associated vectors $\mathbf{Y}_i$, $i = 1, \ldots, M$.

PCA is able to calculate both d – the necessary number of members in the basis of the reduced-order subspace – and the transformation matrix $\mathbf{W}$. PCA proceeds by guaranteeing maximal preserved variance and de-correlation in the latent variable set. From a statistical viewpoint, therefore, it can be assumed that the latent variables are uncorrelated (no linear dependencies among them) or mutually orthogonal, thus constituting a basis. In practice, this means that the covariance matrix of $\mathbb{Y}$, defined as:

$$\mathbf{C}_{yy} = \mathrm{E}\{\mathbb{Y}\mathbb{Y}^T\}, \quad [9.9]$$

is diagonal. Thus, we consider:

$$\begin{aligned} \mathbf{C}_{xx} &= \mathrm{E}\{\mathbb{X}\mathbb{X}^T\} = \mathrm{E}\{\mathbf{W}\mathbb{Y}\mathbb{Y}^T\mathbf{W}^T\} \\ &= \mathbf{W}\mathrm{E}\{\mathbb{Y}\mathbb{Y}^T\}\mathbf{W}^T = \mathbf{W}\mathbf{C}_{yy}\mathbf{W}^T, \end{aligned} \quad [9.10]$$

that by pre-multiplying and post-multiplying by $\mathbf{W}^T$ and $\mathbf{W}$, respectively, and taking into account that $\mathbf{W}^T\mathbf{W} = \mathbf{I}$ leads to:

$$\mathbf{C}_{yy} = \mathbf{W}^T\mathbf{C}_{xx}\mathbf{W}. \quad [9.11]$$

The covariance matrix $\mathbf{C}_{xx}$ can then be factorized by applying the singular-value decomposition:

$$\mathbf{C}_{xx} = \mathbf{V}\mathbf{\Lambda}\mathbf{V}^T, \quad [9.12]$$

with $\mathbf{V}$ containing the orthonormal eigenvectors and $\mathbf{\Lambda}$ the diagonal matrix containing the eigenvalues (non-negative real numbers), assumed in descending order.

Substituting the factorized expression of the covariance matrix [9.12] into equation [9.11], we obtain:

$$\mathbf{C}_{yy} = \mathbf{W}^T\mathbf{V}\mathbf{\Lambda}\mathbf{V}^T\mathbf{W}. \quad [9.13]$$

This equality holds only when the d columns of $\mathbf{W}$ are taken collinear with d columns of $\mathbf{V}$.

From a geometrical viewpoint, the columns of $\mathbf{V}$ indicate the directions in $\mathbb{R}^D$ that span the subspace of the latent variables. We illustrate this interpretation in Figure 9.1, on the left-hand of which, we show points that apparently belong to $\mathbb{R}^2$; however, it is easy to see that all these points belong to a slow one-dimensional manifold. PCA finds an alternative coordinate system given by $\mathbf{V}$ (axes in red), in which all these points are described from a single coordinate.

Figure 9.1. *Geometrical interpretation of PCA. For a color version of the figure, see www.iste.co.uk/brancherie/microstructure.zip*

Nonlinear methods are often more powerful than linear ones, because the connection between the latent variables and the observed ones may be much richer than a simple matrix multiplication. This situation is shown in Figure 9.2, in which it can be noted that no-rotation allows to extract the one-dimensional slow manifold. Thus, PCA indicates that the different points belong to a two-dimensional space, with the risk of concluding that the closest point (using the 2D Euclidean distance) to the red point is in fact the one that is very far from it when using the more appropriate geodesic distance on the one-dimensional slow manifold. Thus, the extraction of the slow manifold is compulsory and PCA is unable to achieve it.

These limitations justify the use of nonlinear dimensionality reduction techniques, such as the local PCA (lPCA), the kernel PCA (kPCA) or the locally linear embedding (LLE).

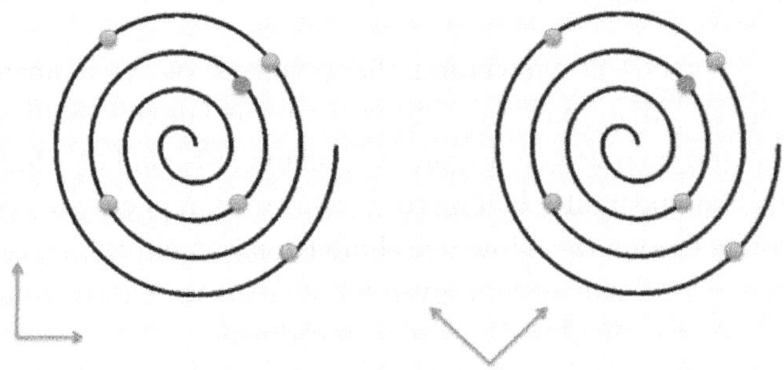

Figure 9.2. *PCA limits in the presence of strongly nonlinear manifolds*

Local PCA applies standard PCA locally, that is, at each data point and its closest neighbors. This is shown in Figure 9.3. The main issue related to its practical implementation is the alignment of the local bases unfolding the slow manifold, as discussed in many papers (for example, [ZHA 04]).

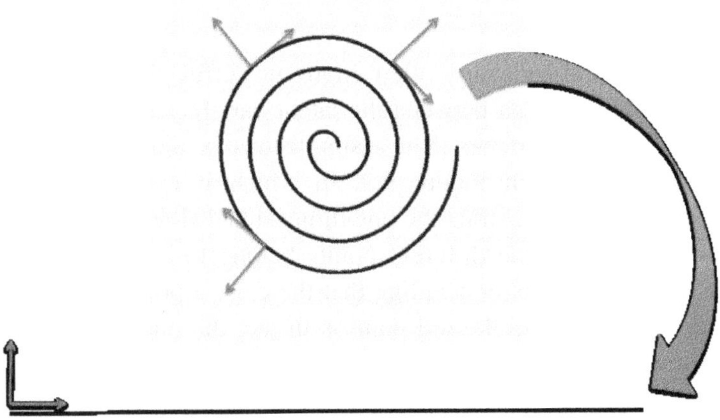

Figure 9.3. *Sketch of local PCA*

9.2.2. *Kernel principal component analysis (kPCA)*

PCA works with the sample covariance matrix, $\mathbb{X}\mathbb{X}^T$. On the contrary, kPCA works with the matrix of pairwise scalar products that defines the Gram matrix $\mathbf{S} = \mathbb{X}^T\mathbb{X}$, as it is also the case of multi-dimensional scaling (MDS) [LEE 07]. In its classical version, MDS preserves pairwise scalar products instead of pairwise distances (both are closely related). It proceeds from:

$$\mathbf{S} = \mathbb{X}^T\mathbb{X} = \mathbb{Y}^T\mathbf{W}^T\mathbf{W}\mathbb{Y} = \mathbb{Y}^T\mathbb{Y}, \qquad [9.14]$$

whose eigenvalue decomposition results:

$$\mathbf{S} = \mathbf{U}\boldsymbol{\Lambda}\mathbf{U}^T = \left(\mathbf{U}\boldsymbol{\Lambda}^{1/2}\right)\left(\boldsymbol{\Lambda}^{1/2}\mathbf{U}^T\right)$$
$$= \left(\boldsymbol{\Lambda}^{1/2}\mathbf{U}^T\right)^T\left(\boldsymbol{\Lambda}^{1/2}\mathbf{U}^T\right),$$

from which it results:

$$\mathbb{Y} = \mathbf{I}_{d\times M}\boldsymbol{\Lambda}^{1/2}\mathbf{U}^T. \qquad [9.15]$$

The idea behind kernel PCA methods is simple: data that are not linearly separable in D dimensions could be done so if previously projected to a space in $Q > D$ dimensions. Thus, surprisingly, kPCA begins by projecting the data to an even higher dimensional space. One of the biggest advantages of this technique is that there is no need to explicitly determine the analytical expression of the mapping.

The symmetric matrix related to the mapped snapshots $\mathbb{Z}^T\mathbb{Z}$ has to be decomposed in eigenvalues and eigenvectors, after performing in the vectors involved in $\mathbb{Z}$, the double centering [LEE 07].

Now, the eigenvalue–eigenvector decomposition can be performed according to:

$$\mathbb{Z}^T\mathbb{Z} = \mathbf{U}\boldsymbol{\Lambda}\mathbf{U}^T, \qquad [9.16]$$

from which it results:

$$\mathbb{Y} = \mathbf{I}_{d \times M} \mathbf{\Lambda}^{1/2} \mathbf{U}^T. \qquad [9.17]$$

It is worth noting that the previous procedure only needs scalar products in the intermediate space $\mathbb{R}^Q$. Mercer's theorem allows to compute such scalar products in the original space $\mathbb{R}^D$, using the so-called kernel trick. There are many possible choices [LEE 07], and then the model calibration becomes its principal advantage as well as its main drawback.

9.2.3. *Locally linear embedding (LLE)*

First we assume the existence of M multi-dimensional data $\mathbf{X}_m$, $m = 1, \ldots, M$, defined in a space of dimension D, that is, $\mathbf{X}_m \in \mathbb{R}^D$. LLE proceeds as follows [ROW 00]:

– Each point $\mathbf{X}_m$, $m = 1, \ldots, M$ is linearly reconstructed from its K-nearest neighbors. In principle, K should be greater than the expected dimension d of the underlying embedded slow manifold and the points should be close enough to ensure the validity of the linear approximation. In general, a sufficiently large number of neighbors K and an appreciably large sampling M ensure a satisfactory reconstruction. For each point $\mathbf{X}_m$, we can write the locally linear data reconstruction as:

$$\mathbf{X}_m = \sum_{i \in \mathcal{S}_m} W_{mi} \mathbf{X}_i, \qquad [9.18]$$

where W_{mi} are the unknown weights and $\mathcal{S}_m$ is the set of the K-nearest neighbors of $\mathbf{X}_m$. As the same weights appear in different locally linear reconstructions, the best compromise is found by looking for the weights, all of them grouped in vector $\mathbf{W}$, that minimize the functional:

$$\mathcal{F}(\mathbf{W}) = \sum_{m=1}^{M} \left\| \mathbf{X}_m - \sum_{i=1}^{M} W_{mi} \mathbf{X}_i \right\|^2 \qquad [9.19]$$

where W_{mi} is zero if $\mathbf{X}_i$ does not belong to the set of K-nearest neighbors of $\mathbf{X}_m$. The minimization of $\mathcal{F}(\mathbf{W})$ allows us to determine all the weights involved in all the locally linear data reconstructions.

– We suppose now that each linear patch around $\mathbf{X}_m$, $\forall m$, is mapped into a lower-dimensional embedding space of dimension d, $d \ll D$. Because of the linear mapping of each patch, weights remain unchanged. The problem becomes the determination of the coordinates of each point $\mathbf{X}_m$ when it is mapped into the low-dimensional space, $\mathbf{Y}_m \in \mathbb{R}^d$.

For this purpose, a new functional $\mathcal{G}$ is introduced, which depends on the searched coordinates $\mathbf{Y}_1, \ldots, \mathbf{Y}_M$:

$$\mathcal{G}(\mathbf{Y}_1, \ldots, \mathbf{Y}_M) = \sum_{m=1}^{M} \left\| \mathbf{Y}_m - \sum_{i=1}^{M} W_{mi} \mathbf{Y}_i \right\|^2, \qquad [9.20]$$

where the weights are known and the reduced coordinates $\mathbf{Y}_m$ are unknown. The minimization of functional $\mathcal{G}$ results in an $M \times M$ eigenvalue problem whose d-bottom non-zero lowest eigenvalues define the set of orthogonal coordinates in which the manifold is mapped.

The use of LLE has some weakness, the first related to the use of Euclidian distances, even if other choices could be considered, and the second is related to a covariance normalization considered when solving the second eigenproblem above. Alternatives, such as the t-SNE [MAA 08], allow us to circumvent the last difficulty.

9.2.4. *Discussion*

The main advantage of local PCA is that it allows for the extraction of the real local reduced dimensionality and the fact that it has a real geometrical transformation allowing not only for extracting the embedded manifold but also to map points outside the slow manifold. Moreover, PCA-based transformations preserve distances, whereas other nonlinear dimensionality reduction strategies fail to accomplish it.

9.3. Manifold-learning-based computational mechanics

Let us assume different physical systems characterized by vectors $\mathbf{X}_m \in \mathbb{R}^D$, whose associated solutions (of the physical problem at hand) are dented by $\mathbf{T}_m$. By applying the LLE (among other possible choices), the slow manifold is extracted. Now, when considering a new physical system, characterized by $\mathbf{X}$, by applying the LLE, its image $\mathbf{Y}$ results on the manifold, with the weights associated to its reconstruction, that is:

$$\mathbf{Y} = \sum_{i \in \mathcal{S}(\mathbf{X})} W_i \mathbf{Y}_i, \qquad [9.21]$$

where $\mathcal{S}(\mathbf{X})$ represents the set of K-nearest neighbors of $\mathbf{X}$. Now, a prediction of the problem solution $\mathbf{T}$ is:

$$\mathbf{T} = \sum_{i \in \mathcal{S}(\mathbf{X})} W_i \mathbf{T}_i. \qquad [9.22]$$

This strategy was successfully considered in [LOP 16b] for addressing models involving parametrized microstructures and shapes. However, there is a strong assumption in the rationale just described. The neighbors and their associated weights in the parametric space are considered to interpolate the solution.

A more accurate approach consists of calculating the parametric solution within, for example, the PGD framework:

$$T(\mathbf{x}, t, y_1, \cdots, y_d) \approx \sum_{i=1}^{N} \mathcal{X}_i(\mathbf{x}) \mathcal{T}_i(t) \mathcal{Y}_i^1(y_1) \cdots \mathcal{Y}_i^d(y_d), \qquad [9.23]$$

where $\mathbf{x}$ denotes the space coordinates involved in usual models and their associated partial differential equations, t is the time involved in transient models and y_j are the latent variables grouped in vector $\mathbf{Y}$ (defining the slow manifold). This procedure was successfully applied in [GON 16] for addressing the same problems that were addressed in [LOP 16b].

9.4. Data-driven simulations

We consider mechanical tests conducted on a perfectly elastic material, in a specimen exhibiting uniform stresses and strains. We will not consider issues related to data generation and collection here; they will be addressed later. More complex behaviors are addressed in [IBA 16]. Thus, for M randomly applied external loads, we assume ourselves able to collect M couples $(\boldsymbol{\sigma}_m, \boldsymbol{\epsilon}_m)$, $m = 1, \ldots, M$. Each stress–strain couple could be represented as a single point $\mathbf{P}_m$ in a phase space of dimension $D = 12$ (six distinct components of the stress and strain tensors). In the following, Voigt notation will be considered, that is, stress and strain tensors will be represented as vectors and consequently the fourth-order elastic tensor reduces to a 6×6 square matrix.

Each vector $\mathbf{P}_m$ thus defines a point in a space of dimension D and, therefore, the whole set of samples represents a set of M points in $\mathbb{R}^D$. We conjecture that all these points belong to (or can be embedded into) a certain low-dimensional manifold embedded into the high-dimensional space $\mathbb{R}^D$ allowing for a nonlinear dimensionality reduction as discussed in [IBA 16]. In the following, we proceed without such a dimensionality reduction and consider the simplest strategy proposed and discussed in [IBA 16]. We consider locally linear approximations, which allow writing:

$$\mathbf{P}_m = \sum_{i=1}^{M} W_{mi} \mathbf{P}_i, \qquad [9.24]$$

with $W_{mi} = 0$ if $i \notin \mathcal{S}_m$ (set containing the K-nearest neighbors of $\mathbf{P}_m$). By minimizing the functional:

$$\mathcal{H}(\mathbf{C}) = \sum_{i \in \mathcal{S}_m} (\boldsymbol{\sigma}_i - \mathbf{C}\boldsymbol{\epsilon}_i)^2, \qquad [9.25]$$

we obtain the secant elastic behavior $\mathbf{C}(\mathbf{P}_m) \equiv \mathbf{C}_m$.

9.4.1. *Data-based weak form*

From the locally linear behavior $\mathbf{C}(\mathbf{P})$ identified above, we could apply the simplest linearization technique operating on the standard weak form:

$$\int_{\Omega} \boldsymbol{\epsilon}^*(\mathbf{x}) : \boldsymbol{\sigma}(\mathbf{x})\, d\mathbf{x} = \int_{\Gamma_N} \mathbf{u}^*(\mathbf{x}) \cdot \mathbf{t}(\mathbf{x})\, d\mathbf{x}, \qquad [9.26]$$

where at each point, from the stress–strain couple at position $\mathbf{x}$, $\mathbf{P}(\mathbf{x})$, the locally linear behavior $\mathbf{C}(\mathbf{P}(\mathbf{x}))$ can be obtained (in practice, at the Gauss points used for the integration of the weak form), which allows us to write, using Voigt notation:

$$\int_{\Omega} \boldsymbol{\epsilon}^*(\mathbf{x}) \cdot (\mathbf{C}(\mathbf{x})\boldsymbol{\epsilon}(\mathbf{x}))\, d\mathbf{x} = \int_{\Gamma_N} \mathbf{u}^*(\mathbf{x}) \cdot \mathbf{t}(\mathbf{x})\, d\mathbf{x}. \qquad [9.27]$$

This allows, in turn, to compute the displacement field and from it to update the strain and stress fields and compute again the locally linear behavior. The process continues until convergence. The discretization related to other two alternative descriptions was considered in depth in [IBA 16].

9.4.2. *Constructing the constitutive manifold*

In the following, we consider perfectly elastic isotropic behavior (that is, inelastic (irreversible) deformations are neglected), and also assume small displacements and deformations. We consider the mechanical specimen occupying the domain $\Omega \in \mathbb{R}^3$ of boundary $\Gamma \equiv \partial\Omega$ with prescribed displacements on Γ_D, $\mathbf{u}(\mathbf{x} \in \Gamma_D) = \mathbf{u}_g$, without loss of generality assumed vanishing, that is, $\mathbf{u}_g = \mathbf{0}$, and prescribe tractions in the complementary boundary Γ_N ($\Gamma = \Gamma_D \cup \Gamma_N$), $\boldsymbol{\sigma} \cdot \mathbf{n}|_{\mathbf{x}\in\Gamma_N} = \mathbf{t}$. Boundary Γ_N is at its turn decomposed in the free-traction region Γ_N^f, where $\mathbf{t} = \mathbf{0}$ and the remaining part Γ_N^t where non-null external tractions, $\mathbf{t} \neq \mathbf{0}$, apply.

We also ignore mass and inertia contributions to the mechanical state and assume that in the absence of external tractions, that is, when $\mathbf{t} = \mathbf{0}$, the mechanical part remains free of strains and stresses, that is, $\boldsymbol{\epsilon} = \mathbf{0}$ and $\boldsymbol{\sigma} = \mathbf{0}$, respectively.

9.4.2.1. *Linear elastic behavior*

The external traction $\mathbf{t}$ is applied, from which the equilibrium reads:

$$\int_{\Omega} \boldsymbol{\epsilon}^*(\mathbf{x}) \cdot \boldsymbol{\sigma}(\mathbf{x}) \, d\mathbf{x} = \int_{\Gamma_N} \mathbf{u}^*(\mathbf{x}) \cdot \mathbf{t} \, d\mathbf{x}. \qquad [9.28]$$

Taking into account the strain- and stress-free reference configuration previously discussed, the problem can be expressed in the incremental form:

$$\int_{\Omega} \Delta\boldsymbol{\epsilon}^*(\mathbf{x}) \cdot \Delta\boldsymbol{\sigma}(\mathbf{x}) \, d\mathbf{x} = \int_{\Gamma_N} \Delta\mathbf{u}^*(\mathbf{x}) \cdot \mathbf{t} \, d\mathbf{x}, \qquad [9.29]$$

or by introducing the linear behavior:

$$\int_{\Omega} \Delta\boldsymbol{\epsilon}^*(\mathbf{x}) \cdot (\mathbf{C}\Delta\boldsymbol{\epsilon}(\mathbf{x})) \, d\mathbf{x} = \int_{\Gamma_N} \Delta\mathbf{u}^*(\mathbf{x}) \cdot \mathbf{t} \, d\mathbf{x}, \qquad [9.30]$$

where the tangent matrix $\mathbf{C}$ is unknown because the linear elastic behavior remains constant throughout the domain. Using a parametrization of symmetric 6×6 matrices (the more general one making use of canonical matrices with a single non-zero entry, taking a unit value), we can write:

$$\mathbf{C} = \sum_{i=1}^{\mathcal{K}} \alpha_i \mathbf{K}_i, \qquad [9.31]$$

with coefficients α_i unknown. When using canonical matrices $\mathcal{K} = 21$, any symmetric matrix can be written from a linear combination of those 21 canonical matrices by considering adequate coefficients α_i, $i = 1, \cdots, 21$.

By introducing this tangent matrix representation into the equilibrium weak form, it results in:

$$\int_{\Omega} \Delta\epsilon^*(\mathbf{x}) \cdot \left(\left(\sum_{i=1}^{\mathcal{K}} \alpha_i \mathbf{M}_i \right) \Delta\epsilon(\mathbf{x}) \right) d\mathbf{x} = \int_{\Gamma_N} \mathbf{\Delta} u^*(\mathbf{x}) \cdot \mathbf{t} \, d\mathbf{x}, \quad [9.32]$$

whose discrete form reads:

$$\Delta\mathbf{U}^* \cdot \left(\sum_{i=1}^{\mathcal{K}} \alpha_i \mathbf{K}_i \right) \Delta\mathbf{U} = \Delta\mathbf{U}^* \cdot \mathbf{T}, \quad [9.33]$$

where $\mathbf{K}_i$ are the rigidity matrices corresponding to the canonical behaviors.

We assume that local displacement and their associated strains are accessible on a certain region of the domain, whose associated degrees of freedom are indicated with the superscript $\bullet^{\mathcal{O}}$. Thus, making use of a partition of the displacements $\Delta\mathbf{U}^{\mathcal{O}}$ and $\Delta\mathbf{U}^{\mathcal{H}}$, referring to the observable and hidden displacements, respectively, the previous discrete nonlinear system can be solved to compute the unknown displacements $\Delta\mathbf{U}^{\mathcal{H}}$ and material coefficients α_i.

When considering a linear behavior, the resulting displacements, strains and stresses can be easily derived from $\mathbf{U} \equiv \Delta\mathbf{U} = (\Delta\mathbf{U}^{\mathcal{H}}, \Delta\mathbf{U}^{\mathcal{O}})^T$ by considering:

$$\begin{cases} \mathbf{u}(\mathbf{x}) = \sum_{i=1}^{\mathcal{N}} \mathbf{U}_i N_i(\mathbf{x}) \\ \epsilon = \nabla_S \mathbf{u} \\ \sigma = \mathbf{C}\epsilon \end{cases}, \quad [9.34]$$

where $\mathcal{N}$ is the number of nodes considered for approximating the displacement field $\mathbf{u}(\mathbf{x})$ and $N_i(\mathbf{x})$ are the associated shape functions.

$\nabla_S \bullet$ denotes the symmetric component of the gradient operator and $\mathbf{C}$ results from α_i:

$$\mathbf{C} = \sum_{i=1}^{\mathcal{K}} \alpha_i \mathbf{K}_i. \qquad [9.35]$$

9.4.2.2. *Nonlinear elastic behavior*

In the nonlinear case, when performing homogeneous tests, it is sufficient to apply the external load incrementally, and at each step, with the same behavior throughout the tested coupon, the behavior will be identified incrementally. However, such a procedure does not allow for exploring the whole strain–stress space, and no testing facilities are able to prescribe any complex multi-axial strain. In our recent works, we proposed an alternative data-driven inverse procedure that, more than using homogeneous tests, exploits complex tests to cover as much as possible the behavioral manifold. This approach combines an incremental loading and a clustering to identify the candidate points, at which the behavior should be updated from the inverse procedure at each loading step.

9.5. Data-driven upscaling of viscous flows in porous media

Isothermal flows of complex fluids in complex microstructures can be simulated by solving the momentum and mass balance equations and a suitable rheological constitutive model. For inertialess incompressible flows, these balance equations read:

$$\nabla \cdot \boldsymbol{\sigma} = \mathbf{0}, \qquad [9.36]$$

and:

$$\nabla \cdot \mathbf{v} = 0, \qquad [9.37]$$

respectively. Here, $\boldsymbol{\sigma}$ is the Cauchy stress tensor and $\mathbf{v}$ is the velocity field, both defined at time t at each point within the fluid domain Ω_f. When considering porous media, the domain Ω is assumed fully saturated, with the fluid phase occupying the region Ω_f, whereas the remaining part $\Omega_s = \Omega - \Omega_f$ is occupied by a solid phase assumed at rest.

An appropriate constitutive equation must be postulated to describe the fluid's rheology. There are many possible choices, the most usual ones being related to Newtonian and generalized Newtonian fluids, as summarized below.

For a Newtonian fluid, the constitutive equation reads:

$$\boldsymbol{\sigma} = -p\mathbf{I} + \boldsymbol{\tau} = -p\mathbf{I} + 2\eta\mathbf{D}, \qquad [9.38]$$

where p is the pressure field that can be interpreted as the Lagrange multiplier associated with the incompressibility constraint, $\mathbf{I}$ is the identity tensor, $\boldsymbol{\tau}$ the extra-stress tensor, η is the constant fluid viscosity and $\mathbf{D}$ is the rate of strain tensor, that is, the symmetric part of the velocity gradient, $2\mathbf{D} = \nabla\mathbf{v} + (\nabla\mathbf{v})^T$.

For a generalized Newtonian fluid, the constitutive equation [9.38] remains formally unchanged, but now the viscosity η depends on the effective strain rate $\dot{\gamma}$ usually expressed from the second invariant of the rate of strain tensor, that is, $\dot{\gamma} = \sqrt{2\mathbf{D} : \mathbf{D}}$. The simplest of such models is the power-law (shear-thinning) viscosity given by:

$$\eta = \kappa\dot{\gamma}^{n-1}, \qquad [9.39]$$

where κ and n are the consistency and power-law index, respectively. The value $n = 1$ corresponds to a Newtonian fluid.

9.5.1. *Upscaling Newtonian and generalized Newtonian fluids flowing in porous media*

The upscaling procedure has been addressed in detail in our previous works [LOP 16a, LOP 17, AMM 16] to handle Newtonian, generalized Newtonian and quasi-Newtonian fluids.

For all these fluids, the flow model is solved in the representative volume $\omega(\mathbf{X})$ located at position $\mathbf{X} \in \Omega$, where two phases coexist, the fluid phase occupying the domain $\omega_f(\mathbf{X})$ and the solid phase, assumed rigid and at rest, occupying the region $\omega_s(\mathbf{X})$, with $\omega_f(\mathbf{X}) \cup \omega_s(\mathbf{X}) = \omega(\mathbf{X})$ and $\omega_f(\mathbf{X}) \cap \omega_s(\mathbf{X}) = \emptyset$. The flow model consists of the mass and momentum balances complemented by the constitutive equation discussed in the previous section:

$$\begin{cases} \nabla \cdot \boldsymbol{\sigma} = \mathbf{0} \\ \nabla \cdot \mathbf{v} = 0 \\ \boldsymbol{\sigma} = -p\mathbf{I} + 2\eta\mathbf{D} = -p\mathbf{I} + \boldsymbol{\tau} \end{cases}, \qquad [9.40]$$

with the viscosity η constant in the case of Newtonian fluids and depending on the effective strain rate in the case of generalized Newtonian fluids. The flow model described above is complemented with the boundary condition $\mathbf{v}(\mathbf{x} \in \partial\omega(\mathbf{X})) = \mathbf{V}$, where $\mathbf{V}$ comes from the macroscopic flow problem.

The solution of the flow problem [9.40] allows calculating the velocity field $\mathbf{v}(\mathbf{x} \in \omega_f(\mathbf{X}))$, where the strain rate is $\mathbf{D}(\mathbf{x} \in \omega_f(\mathbf{X}))$, the local viscosity is $\eta(\mathbf{x} \in \omega_f(\mathbf{X}))$ and the extra-stress tensor is $\boldsymbol{\tau}(\mathbf{x} \in \omega_f(\mathbf{X}))$, which allow calculating the dissipated power in the RVE, $\mathcal{DP}(\mathbf{V};\mathbf{X})$, associated with the prescribed macroscopic velocity $\mathbf{V}$ on its boundary $\partial\omega$:

$$\mathcal{DP}(\mathbf{V};\mathbf{X}) = \int_{\omega_f(\mathbf{X})} \boldsymbol{\sigma}(\mathbf{x}) : \mathbf{D}(\mathbf{x}) d\mathbf{x}, \qquad [9.41]$$

with the specific microscopic dissipation $\mathcal{DP}^m$ obtained by dividing $\mathcal{DP}$, given by [9.41], by the RVE volume $|\omega(\mathbf{X})|$.

Being purely viscous, the model only involves dissipated power, and consequently, the effective macroscopic model should account for that dissipated power. When considering Darcy's model, the specific macroscopic dissipated power $\mathcal{DP}^M$ reads:

$$\mathcal{DP}^M(\nabla P, \mathbf{V}) = \nabla P \cdot \mathbf{V}, \qquad [9.42]$$

where equating the micro- and macro-dissipations results:

$$\mathcal{DP}^m = \nabla P|_{\mathbf{X}} \cdot \mathbf{V}(\mathbf{X}), \qquad [9.43]$$

or by assuming an effective permeability $\mathbf{K}_{eff}(\mathbf{X})$:

$$\nabla P|_{\mathbf{X}} = \mathbf{K}_{eff}^{-1}(\mathbf{X})\mathbf{V}(\mathbf{X}), \qquad [9.44]$$

from which it finally results:

$$\mathcal{DP}^m(\mathbf{V};\mathbf{X}) = \mathbf{K}_{eff}^{-1}(\mathbf{X}) : (\mathbf{V}(\mathbf{X}) \otimes \mathbf{V}(\mathbf{X})). \qquad [9.45]$$

The final expression constitutes a constructive definition of the effective permeability, calculation of which suffices taking the second derivatives of $\mathcal{DP}^m(\mathbf{V})$ related to the microstructure existing at location $\mathbf{X}$:

$$\mathbf{K}_{eff}^{-1}(\mathbf{X}) = \frac{d^2\mathcal{DP}^m(\mathbf{V};\mathbf{X})}{d\mathbf{V}^2}. \qquad [9.46]$$

In the case of a Newtonian fluid, the velocity, strain rate and stress fields scale linearly with the prescribed velocity on the RVE boundary, and consequently, the dissipated power scales with the square of the velocity, leading to a constant permeability, as discussed in detail in [LOP 16a].

9.6. Conclusions

We are not at the beginning of the end, but at the end of the beginning! Data are expected to enrich modeling approaches and even replace poor models in order to improve the accuracy of prediction. The large amount of collected data, including synthetic data generated from simulations, should perform bridging data, information, knowledge and decision making, operating under real-time constraints.

9.7. Bibliography

[AGU 15] AGUADO J.V., HUERTA A., CHINESTA F. *et al.*, "Real-time monitoring of thermal processes by reduced order modelling", *International Journal for Numerical Methods in Engineering*, vol. 102/5, pp. 991–1017, 2015.

[AGU 17] AGUADO J.V., BORZACCHIELLO D., GHNATIOS C. *et al.*, "A simulation App based on reduced order modeling for manufacturing optimization of composite outlet guide vanes", *Advanced Modeling and Simulation in Engineering Sciences*, vol. 4, no. 1, 2017. doi: 10.1186/s40323-017-0087- y.

[AMM 06] AMMAR A., MOKDAD B., CHINESTA F. *et al.*, "A new family of solvers for some classes of multidimensional partial differential equations encountered in kinetic theory modeling of complex fluids", *Journal of Non-Newtonian Fluid Mechanics*, vol. 139, pp. 153–176, 2006.

[AMM 07] AMMAR A., MOKDAD B., CHINESTA F. *et al.*, "A new family of solvers for some classes of multidimensional partial differential equations encountered in kinetic theory modeling of complex fluids. Part II: Transient simulation using space-time separated representation", *Journal of Non-Newtonian Fluid Mechanics*, vol. 144, pp. 98–121, 2007.

[AMM 10] AMMAR A., CHINESTA F., FALCO A., "On the convergence of a greedy rank-one update algorithm for a class of linear systems", *Archives of Computational Methods in Engineering*, vol. 17/4, pp. 473–486, 2010.

[AMM 11] AMMAR A., CHINESTA F., CUETO E. *et al.*, "Proper generalized decomposition of time-multiscale models", *International Journal for Numerical Methods in Engineering*, vol. 90/5, pp. 569–596, 2011.

[AMM 14] AMMAR A., HUERTA A., CHINESTA F. *et al.*, "Parametric solutions involving geometry: a step towards efficient shape optimization", *Computer Methods in Applied Mechanics and Engineering*, vol. 268C, pp. 178–193, 2014.

[AMM 16] AMMAR A., ABISSET-CHAVANNE E., CHINESTA F. *et al.*, "Flow modelling of quasi-Newtonian fluids in two-scale fibrous fabrics. Advanced simulations", *International Journal of Material Forming*, in press. doi: 10.1007/s12289-016-1300-0.

[BAR 14] BARBARULO A., RIOU H., KOVALEVSKY L. *et al.*, "PGD-VTCR: a reduced order model technique to solve medium frequency broad band problems on complex acoustical systems", *Journal of Mechanical Engineering*, vol. 60/5, pp. 307–314, 2014.

[BOG 12] BOGNET B., LEYGUE A., CHINESTA F. *et al.*, "Advanced simulation of models defined in plate geometries: 3D solutions with 2D computational complexity", *Computer Methods in Applied Mechanics and Engineering*, vol. 201, pp. 1–12, 2012.

[BOG 14] BOGNET B., LEYGUE A., CHINESTA F., "Separated representations of 3D elastic solutions in shell geometries", *Advanced Modelling and Simulation in Engineering Sciences*, vol. 1, no. 4, available at: http://www.amses-journal.com/content/1/1/4, 2014.

[BOR 15] BORDEU F., GHNATIOS CH., BOULZE D. *et al.*, "Parametric 3D elastic solutions of beams involved in frame structures", *Advances in Aircraft and Spacecraft Science*, vol. 2/3, pp. 233–248, 2015.

[BOR 17] BORZACCHIELLO D., AGUADO J.V., CHINESTA F., "Reduced order modelling for efficient numerical optimisation of a hot-wall chemical vapour deposition reactor", *International Journal of Numerical Methods for Heat and Fluid Flow*, vol. 27, no. 7, pp. 1602–1622, 2017. doi: 10.1108/HFF-04-2016-0153.

[BOU 97] BOUCARD P.A., LADEVEZE P., POSS M. *et al.*, "A non-incremental approach for large displacement problems", *Computers & Structures*, vol. 64, pp. 499–508, 1997.

[BOU 13] BOUCINHA L., GRAVOUIL A., AMMAR A., "Space-time proper generalized decompositions for the resolution of transient elastodynamic models", *Computer Methods in Applied Mechanics and Engineering*, vol. 255, pp. 67–88, 2013.

[CHI 10] CHINESTA F., AMMAR A., CUETO E., "Recent advances and new challenges in the use of the proper generalized decomposition for solving multidimensional models", *Archives of Computational Methods in Engineering*, vol. 17/4, pp. 327–350, 2010.

[CHI 11] CHINESTA F., LADEVEZE P., CUETO E., "A short review in model order reduction based on proper generalized decomposition", *Archives of Computational Methods in Engineering*, vol. 18, pp. 395–404, 2011.

[CHI 13a] CHINESTA F., KEUNINGS R., LEYGUE A., *The Proper Generalized Decomposition for Advanced Numerical Simulations. A Primer*, Springer, 2013.

[CHI 13b] CHINESTA F., LEYGUE A., BORDEU F. *et al.*, "Parametric PGD based computational vademecum for efficient design, optimization and control", *Archives of Computational Methods in Engineering*, vol. 20/1, pp. 31–59, 2013.

[CHI 14] CHINESTA F., LEYGUE A., BOGNET B. *et al.*, "First steps towards an advanced simulation of composites manufacturing by automated tape placement", *International Journal of Material Forming*, vol. 7/1, pp. 81–92, 2014.

[CHI 16] CHINESTA F., HUERTA A., ROZZA G. *et al.*, *Model Order Reduction*, 2nd edition, Wiley, 2016.

[CRE 13] CREMONESI M., NERON D., GUIDAULT P.A. *et al.*, "A PGD-based homogenization technique for the resolution of nonlinear multiscale problems", *Computer Methods in Applied Mechanics and Engineering*, vol. 267, pp. 275–292, 2013.

[GHN 11] GHNATIOS CH., CHINESTA F., CUETO E. *et al.*, "Methodological approach to efficient modeling and optimization of thermal processes taking place in a die: application to pultrusion", *Composites Part A*, vol. 42, pp. 1169–1178, 2011.

[GHN 12] GHNATIOS CH., MASSON F., HUERTA A. *et al.*, "Proper generalized decomposition based dynamic data-driven control of thermal processes", *Computer Methods in Applied Mechanics and Engineering*, vol. 213, pp. 29–41, 2012.

[GON 12] GONZALEZ D., MASSON F., POULHAON F. *et al.*, "Proper generalized decomposition based dynamic data-driven inverse identification", *Mathematics and Computers in Simulation*, vol. 82/9, pp. 1677–1695, 2012.

[GON 14] GONZALEZ D., CUETO E., CHINESTA F., "Real-time direct integration of reduced solid dynamics equations", *International Journal for Numerical Methods in Engineering*, vol. 99/9, pp. 633–653, 2014.

[GON 15] GONZALEZ D., ALFARO I., QUESADA C. *et al.*, "Computational vademecums for the real-time simulation of haptic collision between nonlinear solids", *Computer Methods in Applied Mechanics and Engineering*, vol. 283, pp. 210–223, 2015.

[GON 16] GONZALEZ D., AGUADO J.V., ABISSET-CHAVANNE E. *et al.*, "kPCA-based parametric solutions within the PGD framework", *Archives of Computational Methods in Engineering*. doi: 10.1007/s11831-016-9173-4, 2016.

[HEY 13] HEYBERGER CH., BOUCARD P.A., NERON D., "A rational strategy for the resolution of parametrized problems in the PGD framework", *Computer Methods in Applied Mechanics and Engineering*, vol. 259, pp. 40–49, 2013.

[IBA 16] IBANEZ R., ABISSET-CHAVANNE E., AGUADO J.V. *et al.*, "A manifold-based methodological approach to data-driven computational elasticity and inelasticity", *Archives of Computational Methods in Engineering*. doi: 10.1007/s11831-016-9197-9, 2016.

[LAD 85] LADEVEZE P., "On a family of algorithms for structural mechanics (in french)", *Comptes Rendus Académie des Sciences Paris*, vol. 300/2, pp. 41–44, 1985.

[LAD 89] LADEVEZE P., "The large time increment method for the analyze of structures with nonlinear constitutive relation described by internal variables", *Comptes Rendus Académie des Sciences Paris*, vol. 309, pp. 1095–1099, 1989.

[LAD 96] LADEVEZE P., *Mécanique non linéaire des structures*, Hermes, Paris, 1996.

[LAD 99] LADEVEZE P., *Nonlinear Computational Structural Mechanics. New Approaches and Non-incremental Methods of Calculation*, Springer Verlag, 1999.

[LAM 10] LAMARI H., AMMAR A., CARTRAUD P. *et al.*, "Routes for efficient computational homogenization of nonlinear materials using the proper generalized decomposition", *Archives of Computational Methods in Engineering*, vol. 17/4, pp. 373–391, 2010.

[LEE 07] LEE J.A., VERLEYSEN M., *Nonlinear Dimensionality Reduction*, Springer, Berlin, 2007.

[LOP 16a] LOPEZ E., ABISSET-CHAVANNE E., COMAS-CARDONA S. *et al.*, "Flow modeling of linear and nonlinear fluids in two and three scale fibrous fabrics", *International Journal of Material Forming*, pp. 215–227, doi: 10.1007/s12289-015-1224-0.

[LOP 16b] LOPEZ E., GONZALEZ D., ABISSET-CHAVANNE E. *et al.*, "A manifold learning approach for integrated computational materials engineering", *Archives of Computational Methods in Engineering*. doi: 10.1007/s11831-016-9172-5, 2016.

[LOP 17] LOPEZ E., LEYGUE A., ABISSET-CHAVANNE E. *et al.*, "Flow modeling of linear and nonlinear fluids in two scale fibrous fabrics. Advanced simulations", *International Journal of Material Forming*, pp. 317–328, doi: 10.1007/s12289-015-1280-5, 2017.

[MAA 08] VAN DER MAATEN L.J.P., HINTON G.E., "Visualizing high-dimensional data using t-SNE", *Journal of Machine Learning Research*, vol. 9, pp. 2579–2605, 2008.

[NER 10] NÉRON D., LADEVÈZE P., "Proper generalized decomposition for multiscale and multiphysics problems", *Archives of Computational Methods in Engineering*, vol. 17/4, pp. 351–372, 2010.

[NER 15] NÉRON D., BOUCARD P.A., RELUN N., "Time-space PGD for the rapid solution of 3D nonlinear parametrized problems in the many-query context", *International Journal for Numerical Methods in Engineering*, vol. 103/4, pp. 275–292, 2015.

[NOU 08] NOUY A., "Generalized spectral decomposition method for solving stochastic finite element equations: invariant subspace problem and dedicated algorithms", *Computer Methods in Applied Mechanics and Engineering*, vol. 197, pp. 4718–4736, 2008.

[NOU 09a] NOUY A., LE MAITRE O., "Generalized spectral decomposition method for stochastic nonlinear problems", *Journal of Computational Physics*, vol. 228/1, pp. 202–235, 2009.

[NOU 09b] NOUY A., "Recent developments in spectral stochastic methods for the numerical solution of stochastic partial differential equations", *Archives of Computational Methods in Engineering*, vol. 16/3, pp. 251–285, 2009.

[NOU 10] NOUY A., "Proper generalized decompositions and separated representations for the numerical solution of high dimensional stochastic problems", *Archives of Computational Methods in Engineering – State of the Art Reviews*, vol. 17, pp. 403–434, 2010.

[RIO 13] RIOU H., LADEVEZE P., KOVALEVSKY L., "The variational theory of complex rays: an answer to the resolution of mid-frequency 3D engineering problems", *Journal of Sound and Vibration*, vol. 332, pp. 1947–1960, 2013.

[ROW 00] ROWEIS S.T., SAUL L.K., "Nonlinear dimensionality reduction by locally linear embedding", *Science*, vol. 290/5500, pp. 2323–2326, 2000.

[VID 12] VIDAL P., GALLIMARD L., POLIT O., "Composite beam finite element based on the proper generalized decomposition", *Computers & Structures*, vol. 102, pp. 76–86, 2012.

[VID 13] VIDAL P., GALLIMARD L., POLIT O., "Proper generalized decomposition and layer-wise approach for the modeling of composite plate structures", *International Journal of Solids and Structures*, vol. 50/14–15, pp. 2239–2250, 2013.

[ZHA 04] ZHANG Z., ZHA H., Principal manifolds and nonlinear dimensionality reduction via tangent space alignment", *SIAM: SIAM Journal on Scientific Computing*, vol. 26/1, pp. 313–338, 2004.

List of Authors

Emmanuelle Abisset-Chavanne
Ecole Centrale de Nantes
Nantes
France

Jose Vicente Aguado
Ecole Centrale de Nantes
Nantes
France

José Luis Alves
Mechanical Engineering Department
Engineering School
University of Minho
Portugal

Domenico Borzacchiello
Ecole Centrale de Nantes
Nantes
France

Salima Bouvier
Department of Mechanical Engineering
University of Technology of Compiègne
France

Delphine Brancherie
Department of Mechanical Engineering
University of Technology of Compiègne
France

Oana Cazacu
Department of Mechanical and Aerospace Engineering
University of Florida – REEF
Shalimar
United States

Francisco Chinesta
Ecole Centrale de Nantes
Nantes
France

Elias Cueto
University of Zaragoza
Spain

Alberto M. Cuitiño
Rutgers University
New Jersey
United States

Félix Darve
Grenoble Institute of Technology
France

Jérôme Duriez
University of Calgary
Alberta
Canada

Pierre Feissel
Roberval Laboratory
University of Technology of Compiègne
France

Jean-François Ganghoffer
University of Lorraine
Nancy
France

David Gonzalez
University of Zaragoza
Spain

Marcial Gonzalez
Purdue University
Indiana
United States

John Hermanson
Cornell University
Ithaca
United States

Adnan Ibrahimbegović
Roberval Laboratory
University of Technology of Compiègne
France

Athanasios Iliopoulos
United States Naval Research Laboratory
Washington, D.C.
United States

Emir Karavelić
University of Sarajevo
Bosnia and Herzegovina

Foteini Komnineli
Computational Material Science Center
George Mason University
Fairfax
and
United States Naval Research Laboratory
Washington, D.C.
United States

Hassan Lakiss
Department of Engineering
Lebanese University
Beirut
Lebanon

Elena Lopez
Ecole Centrale de Nantes
Nantes
France

John Michopoulos
United States Naval Research Laboratory
Washington, D.C.
United States

Mijo Nikolić
University of Split
Croatia

Adam L. Pilchak
Air Force Research Laboratory
Ohio
United States

Henry Proudhon
Centre des Matériaux
MINES ParisTech
France

Yosra Rahali
University of Lorraine
Nancy
France

Hilal Reda
University of Lorraine
Nancy
France

John Steuben
United States Naval Research Laboratory
Washington, D.C.
United States

Joseph C. Tucker
Exponent
California
United States

Richard Wan
University of Calgary
Alberta
Canada

Tyler J. Weihing
Aerospace Systems Design Laboratory
Georgia Institute of Technology
Atlanta
United States

Bereket Yohannes
Rutgers University
New Jersey
United States

Index

A, B, C

D, E, F

G, H, I

L, M, P

R, S

T, V, W

Other titles from

iSTE

in

Civil Engineering and Geomechanics

2017

LAUZIN Xavier
Civil Engineering Structures According to the Eurocodes

SELLIER Alain, GRIMAL Étienne, MULTON Stéphane, BOURDAROT Éric
Swelling Concrete in Dams and Hydraulic Structures: DSC 2017

2016

BARRE Francis, BISCH Philippe, CHAUVEL Danièle *et al.*
Control of Cracking in Reinforced Concrete Structures: Research Project CEOS.fr

PIJAUDIER-CABOT Gilles, LA BORDERIE Christian, REESS Thierry, CHEN Wen, MAUREL Olivier, REY-BETHBEDER Franck, DE FERRON Antoine
Electrohydraulic Fracturing of Rocks

TORRENTI Jean-Michel, LA TORRE Francesca
Materials and Infrastructures 1
Materials and Infrastructures 2

2015

AÏT-MOKHTAR Abdelkarim, MILLET Olivier
Structure Design and Degradation Mechanisms in Coastal Environments

MONNET Jacques
In Situ Tests in Geotechnical Engineering

2014

DAÏAN Jean-François
Equilibrium and Transfer in Porous Media – 3-volume series
Equilibrium States – Volume 1
Transfer Laws – Volume 2
Applications, Isothermal Transport, Coupled Transfers – Volume 3

2013

AMZIANE Sofiane, ARNAUD Laurent
Bio-aggregate-based Building Materials: Applications to Hemp Concretes

BONELLI Stéphane
Erosion in Geomechanics Applied to Dams and Levees

CASANDJIAN Charles, CHALLAMEL Noël, LANOS Christophe, HELLESLAND Jostein
Reinforced Concrete Beams, Columns and Frames: Mechanics and Design

GUÉGUEN Philippe
Seismic Vulnerability of Structures

HELLESLAND Jostein, CHALLAMEL Noël, CASANDJIAN Charles, LANOS Christophe
Reinforced Concrete Beams, Columns and Frames: Section and Slender Member Analysis

LALOUI Lyesse, DI DONNA Alice
Energy Geostructures: Innovation in Underground Engineering

LEGCHENKO Anatoly
Magnetic Resonance Imaging for Groundwater

2012

BONELLI Stéphane
Erosion of Geomaterials

JACOB Bernard *et al.*
ICWIM6 – Proceedings of the International Conference on Weigh-In-Motion

OLLIVIER Jean-Pierre, TORRENTI Jean-Marc, CARCASSES Myriam
Physical Properties of Concrete and Concrete Constituents

PIJAUDIER-CABOT Gilles, PEREIRA Jean-Michel
Geomechanics in CO_2 Storage Facilities

2011

BAROTH Julien, BREYSSE Denys, SCHOEFS Franck
Construction Reliability: Safety, Variability and Sustainability

CREMONA Christian
Structural Performance: Probability-based Assessment

HICHER Pierre-Yves
Multiscales Geomechanics: From Soil to Engineering Projects

IONESCU Ioan R. *et al.*
Plasticity of Crystalline Materials: from Dislocations to Continuum

LOUKILI Ahmed
Self Compacting Concrete

MOUTON Yves
Organic Materials for Sustainable Construction

NICOT François, LAMBERT Stéphane
Rockfall Engineering

PENSE-LHERITIER Anne-Marie
Formulation

PIJAUDIER-CABOT Gilles, DUFOUR Frédéric
Damage Mechanics of Cementitious Materials and Structures

RADJAI Farhang, DUBOIS Frédéric
Discrete-element Modeling of Granular Materials

RESPLENDINO Jacques, TOUTLEMONDE François
Designing and Building with UHPFRC

2010

ALSHIBLI A. Khalid
Advances in Computed Tomography for Geomechanics

BUZAUD Eric, IONESCU Ioan R., VOYIADJIS Georges
Materials under Extreme Loadings / Application to Penetration and Impact

LALOUI Lyesse
Mechanics of Unsaturated Geomechanics

NOVA Roberto
Soil Mechanics

SCHREFLER Bernard, DELAGE Pierre
Environmental Geomechanics

TORRENTI Jean-Michel, REYNOUARD Jean-Marie, PIJAUDIER-CABOT Gilles
Mechanical Behavior of Concrete

2009

AURIAULT Jean-Louis, BOUTIN Claude, GEINDREAU Christian
Homogenization of Coupled Phenomena in Heterogenous Media

CAMBOU Bernard, JEAN Michel, RADJAI Fahrang
Micromechanics of Granular Materials

MAZARS Jacky, MILLARD Alain
Dynamic Behavior of Concrete and Seismic Engineering

NICOT François, WAN Richard
Micromechanics of Failure in Granular Geomechanics

2008

BETBEDER-MATIBET Jacques
Seismic Engineering

CAZACU Oana
Multiscale Modeling of Heterogenous Materials

HICHER Pierre-Yves, SHAO Jian-Fu
Soil and Rock Elastoplasticity

JACOB Bernard *et al.*
HVTT 10

JACOB Bernard *et al.*
ICWIM 5

SHAO Jian-Fu, BURLION Nicolas
GeoProc2008

2006

BALAGEAS Daniel, FRITZEN Claus-Peter, GÜEMES Alfredo
Structural Health Monitoring

DESRUES Jacques *et al.*
Advances in X-ray Tomography for Geomaterials

FSTT
Microtunneling and Horizontal Drilling

MOUTON Yves
Organic Materials in Civil Engineering

2005

PIJAUDIER-CABOT Gilles, GÉRARD Bruno, ACKER Paul
Creep Shrinkage and Durability of Concrete and Concrete Structures CONCREEP – 7

Printed and bound by CPI Group (UK) Ltd, Croydon, CR0 4YY

15/09/2026

14962345-0003